CW01266379

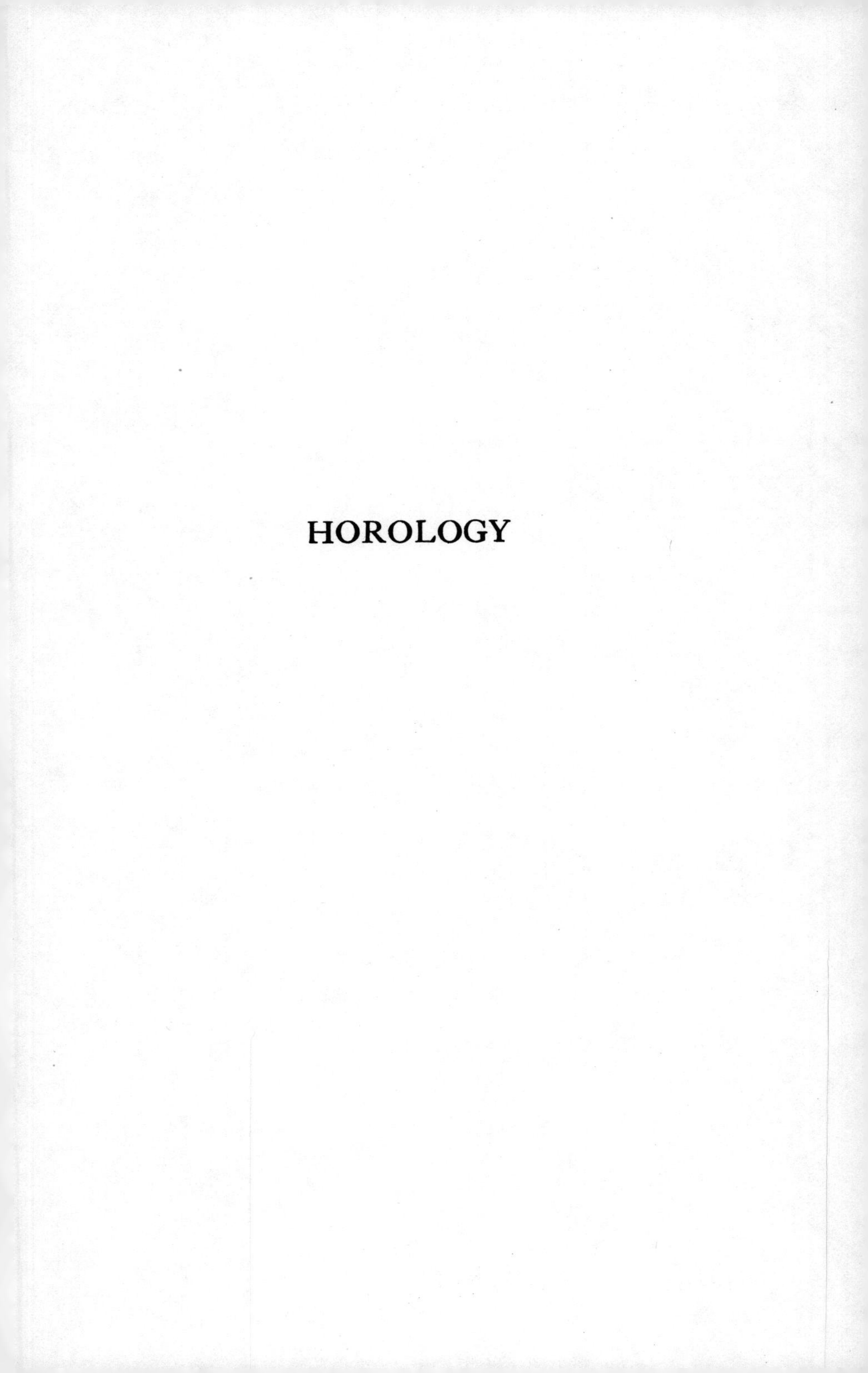

HOROLOGY

HOROLOGY

The Science of Time Measurement and the Construction of Clocks, Watches and Chronometers

BY

J. ERIC HASWELL, F.B.H.I.

SILVER MEDALLIST, CITY AND GUILDS OF LONDON INSTITUTE, 1912
SILVER MEDALLIST, THE WORSHIPFUL COMPANY OF TURNERS, LONDON, 1912
BRONZE MEDALLIST, BRITISH HOROLOGICAL INSTITUTE, 1911
FIRST PRIZEMAN, THE WORSHIPFUL COMPANY OF SKINNERS, LONDON, 1909–10; ETC.

Citizen and Liveryman, The Worshipful Company of Clockmakers, London

WITH SUPPLEMENT

EP PUBLISHING LIMITED
CALDWELL INDUSTRIES
1976

First published 1928 by Chapman and Hall Ltd., London
This edition reprinted from the 1951 edition

Republished 1976 in Great Britain by
EP Publishing Limited
East Ardsley, Wakefield
West Yorkshire, England
and in the United States of America by
Caldwell Industries
Salt Lake City, Utah

by kind permission of the copyright holder

Second impression of this edition, 1976

ISBN 0 7158 1146 0

Please address all enquiries to EP Publishing Limited
(address as above)

Printed in Great Britain by
Redwood Burn Limited
Trowbridge & Esher

TO

M. R. R. H.

PREFACE

In the preparation of this book, the intention has been to explain the principles of operation and the constructional details of clocks, watches and chronometers in such a manner as to assist the student and provide a work of reference for the more experienced. Appropriate historical facts have been embodied wherever desirable, but the primary object of the book is to furnish technical data and information, supplemented by clear and accurate diagrams.

At a glance it is easy to see how greatly diverse are the mechanical features of, for instance, a turret clock and a wrist watch, yet these and other timepieces are all embraced in the same technology.

In addition to the pages relating to the fundamental principles of time measurement, there are those concerning calculations of a simple character for finding lengths of pendulums, numbers for wheels and pinions, etc. Care has also been devoted to such essential details as the various forms of escapements, striking and chiming mechanisms, repeating and chronograph work, balances and springs, keyless devices, as well as electrical clock systems and marine chronometers, accompanying all of which will be found helpful scale drawings and diagrams.

Looking back, now many years, to student days, it is possible to recall pleasant memories of how we young horological aspirants received with reverence and awe the able tuition and guidance of Mr. Thomas D. Wright, whose name will ever remain as that of one of the most prominent authorities in this country on technical horology. Much of Mr. Wright's instruction was accompanied by manuscript notes which subsequently appeared in serial form in the *Horological Journal.* In so far as I have had the rising generation in mind I have, in certain instances, gone back to those class notes for reference, because of their adaptability to the subject.

To my valued friend and late instructor, Mr. Heinrich Otto, I owe a particular debt of gratitude for the exceptional interest, thoroughness and devotion he has expended upon reading the manuscript. His sound advice, wide experience and profound knowledge, so generously placed at my disposal, have proved of inestimable worth to me. I should like to add that in preparing the drawings of escapements I have followed exclusively the method of instruction which was adopted by Mr. Otto, during the time that he held office as Drawing Instructor of the British Horological Institute.

Also to Mr. C. O. Bartrum, B.Sc., of the Hampstead Scientific Society, my thanks are due for his kindness in reading the part of the manuscript dealing with the astronomical aspect of time measurement.

For their help with various details and material to enable the production of certain illustrations, I should like to take this opportunity of expressing my thanks to Mr. J. F. Cole, Messrs. Cooke, Troughton & Simms, Ltd. (Plates I and II), Sir Frank Dyson, Messrs. J. B. Joyce & Co., Ltd. (Plates III, IV, V and VI), Mr. D. J. Parkes and others.

J. Eric Haswell.

Clerkenwell,
London, E.C.
May, 1928.

CONTENTS

PART IV

MARINE CHRONOMETERS

LIST OF PLATES

LIST OF ILLUSTRATIONS

HOROLOGY

PART I—INTRODUCTORY

CHAPTER I

TIME MEASUREMENT

The Earth as a Clock. Horology as a mechanical science embraces not only the essential factors concerning the determination of time but also the principles governing the design of instruments whereby the passage of time is denoted. Time measurement being the primary consideration, it is necessary to have some definite basis to work on—some self-regulating timepiece—some perfect "master" clock. This perfection is found in the rotation of the earth on its axis. The earth itself is a perfect clock —indeed it is the only perfect clock applicable to our purpose—and so far as can be ascertained there has been, from the earliest ages, only an infinitesimal change in the period occupied by its successive rotations.

Determination of this period is rendered possible by a very elementary study of the stars. Excluding the planets which, in common with the earth, perform their own little evolutions round the sun, the myriads of stars visible on a clear night are known as "fixed" stars and form constellations in parts of the heavens inconceivably remote. The effect of these great distances, therefore, makes the relative movement of the earth about the sun unnoticeable to an imaginary observer situated on one of these fixed stars. Reversing the order, an observer on the earth looking through a rigidly planted telescope at one of these stars will see it pass across his field of view and, after the earth has performed one complete rotation, he will see the same star appear again in his field. It is the successive reappearance of such a star which enables one to compre-

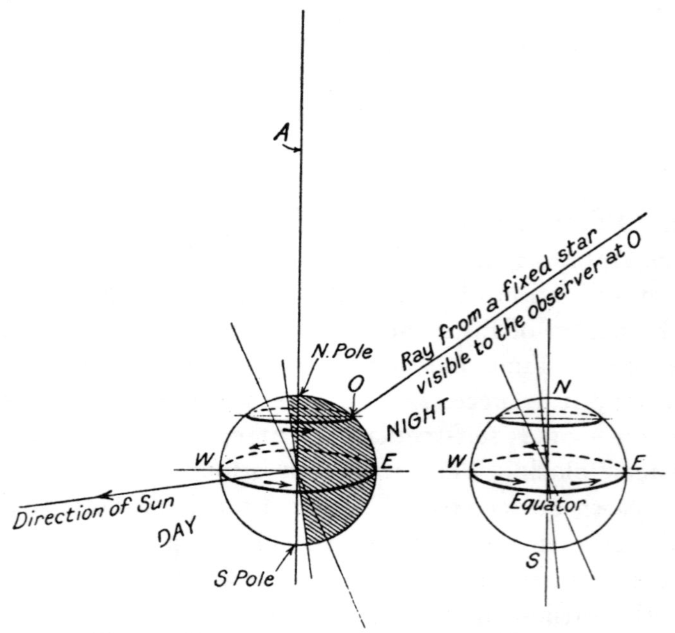

FIG. 1.—Diagram showing the direction of the rotation of the Earth about its own axis.

hend the regularity of the motion of the earth about its own axis and the uniformity of its velocity. It must be remembered that the apparent motion of the stars is really due to the motion of the earth, though for the convenience of observers it is naturally easier to think of the earth as remaining still.

Sidereal Time. Fig. 1 is intended to illustrate the direction of rotation of the earth about its own axis. Consider the ball as representing our earth suspended by a long string, *A*, which enters at the northern end or "pole" of the axis and is, therefore, in the same straight line as the axis. Imagining the ball to revolve from left to right, looking down the axis from *A*, an idea of the earth's rotation, from west to east, can easily be formed. This direction of motion is described as "anti-clockwise." The figure also shows a line encircling the ball to represent the equator, the plane of which, passing through the centre of the earth, is at right angles to the axis. At point *O* an observer is looking through a telescope at some fixed star which is said to "cross his field," though in actual fact it is the movement of the earth which causes his field to move out of the range of the star. The intervening period between two successive appearances of that star in the observer's field is known as one "sidereal" or star day. The path shown on the earth through *O* will have been traversed apparently by the star during that period. A sidereal day thus measured is divided equally into 24 hours, which, in turn, are divided into 60 equal minutes of 60 equal seconds each, comprising the constituents of "Sidereal Time" measurement. This form of time differs from that in general use and is employed almost exclusively by astronomers, as will be seen later.

Solar Time. It is now necessary to turn to the other aspect of the earth's motion. In addition to rotating about its own axis, in an anti-clockwise direction at a uniform rate, the earth also progresses, with varying velocity, in a similar direction, along an enormous elliptical path or orbit around the sun in a year of approximately 365 days. The sun is not situated exactly in the centre of the ellipse but towards one of the foci, and the earth's velocity is greatest when the earth is nearest the sun. The consecutive reappearance of the sun before any part of the earth's surface constitute what are known as "solar" or sun days. The effect, however, of the irregularities of diurnal progress produces solar days of varying lengths, which are consequently useless as the all-important constant standard for time measurement. True sun days are actually recorded on that familiar instrument, the sundial. Nevertheless, though solar observations are useless for determining a constant standard, the fact that the successive reappearance differentiates between periods of darkness and light makes solar time obviously the most convenient form to employ for ordinary or civil purposes.

Mean Solar Time. In actual practice, however, in order to eliminate the irregularities of the solar days and to produce uniformity, an average or mean of these unequal stages of the earth's annual journey is taken, resulting in what are known as "mean solar days." These days are of even duration and, like the sidereal day, are equally divided into 24 hours, each of which is sub-divided into 60 minutes of 60 seconds each. It is important, however, to distinguish clearly in one's mind between these two methods of time measurement, because, although both the sidereal day and the mean solar day are divided into 24 hours, etc., the periods differ somewhat in length. In terms of mean solar time, the sidereal day is 23 hours 56 minutes 4·09 seconds in duration; that is, the sidereal day is shorter than the mean solar day by 3 minutes, 55·91 seconds of mean solar time, which represents a loss on sidereal time of 9·83 seconds per hour. Conversely, in terms of sidereal time, the mean solar day is longer than the sidereal day by 3 minutes 56·55 seconds, representing a gain of 9·8565 seconds per hour on mean solar time.

Fig. 2 shows the sun situated in the left-hand focus of an ellipse which represents the earth's orbit. When drawn to this small scale the orbit is very nearly a circle. The earth itself is indicated at six different stages on its annual journey, the arrows showing the direction of motion. The spheres are marked with meridians which intersect at the north pole above and at the south pole beneath, and are also drawn to show the inclination of the earth to the

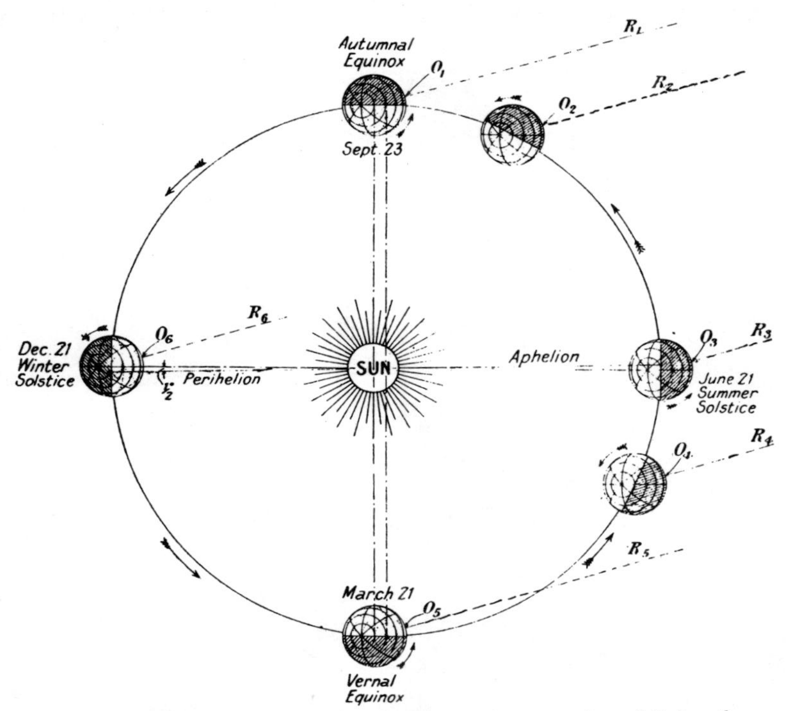

FIG. 2.—Illustrating the passage of the Earth along its orbital path around the Sun.

ecliptic. The small circles centred at the poles are the arctic circle and, in the dotted part, the antarctic circle (cf. Fig. 3). The large ones near the circumference indicate the equator. The positions of the spheres represent the earth at the summer and winter solstices and the vernal and autumnal equinoxes, in addition to two intermediate positions. The shaded portions differentiate between night and day, from which it will be noticed that, at the winter solstice, the part of the earth north of the arctic circle is never in daylight; whilst, at the summer solstice, it is never night in the arctic region. The reverse, of course, applies to the antarctic circle in the southern hemisphere. The earth is nearer the sun when it is winter in the northern

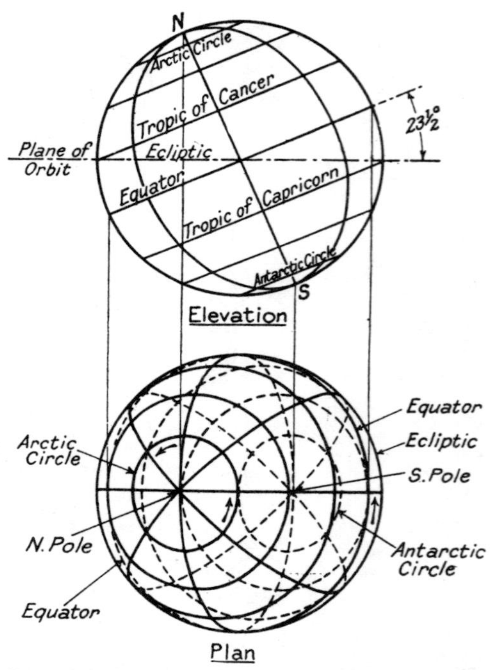

FIG. 3.—The Earth, in Elevation along the plane of the orbit and in Plan, perpendicular to the orbit from above the Northern Hemisphere.

hemisphere, and the shortest distance between the earth and the sun measured along the major axis of the ellipse is called Perihelion, whilst the longest distance is called Aphelion.

Rays from an infinitely distant star approach the earth along R_1, R_2, R_3, R_4, R_5, R_6 (Fig. 2) and, because of the vast distance, the movement of the earth along its orbit makes no apparent difference to the moment this star will successively reappear before a given observer stationed at O_1, O_2, O_3, O_4, O_5, O_6 as the earth rotates on its axis. In the figure, the observer can only see this particular star during summer; in winter, the light of the sun prevents it being visible. This fact should make clear the inconvenience of using sidereal time for civil purposes.

Fig. 3 represents the earth, in elevation along the plane of the orbit, and in plan, which is an enlarged view of the spheres in Fig. 2, as it would appear from above the northern hemisphere and perpendicular to the orbit. The axial inclination of $23\frac{1}{2}^\circ$ (at present actually $23^\circ\ 27'\ 8{\cdot}26''$) from the vertical corresponds to the obliquity of the ecliptic, which is the name given to the apparent path of the sun upon the earth during the year.

Longitude and Latitude. For convenience in locating or describing the geographical position of any place, the earth's surface is dissected by imaginary circles just as the map of a town or district is sometimes divided into half-mile squares and marked off in letters one way and numbers in the other. These lines of demarkation, which represent angular distances radiating from the centre of the earth, run north and south, intersecting at the poles, and east and west parallel to the equator. The former are known as "meridians of longitude" and the latter as "parallels of latitude." The prime meridian is considered to pass through Greenwich, from whence other meridians are marked off in degrees of arc "east" and "west," and any observer's "terrestrial" meridian is that which passes through his own station. The circumference of the earth is, of course, 360 degrees, so that the 180th degree E. (somewhere in the Pacific Ocean) is also the 180th degree W. As time reckonings are almost universally regulated from Greenwich, different places being so much before or after, one can imagine a day as commencing in the Pacific and ending in the Pacific. A line roughly following the 180th meridian, and known as the "date line," has been determined for the purpose of indicating the change from one day to another.

Parallels of latitude are marked off similarly in degrees north and south of the equator, which is zero.

These degrees are, in both cases, subdivided into the usual minutes and seconds of arc, and the position of any place on the earth's surface can be very precisely located when its longitude E. or W. of some prime meridian, such as Greenwich, and latitude N. or S. are known.

It should be noted that the meridians can also be expressed in time, 1 degree being equal to 4 minutes of time. That is arrived at by dividing 360 degrees, the circumference of the earth's surface, by 24 hours, the period of a complete revolution. The time value for longitude is of great importance in making celestial observations.

At the moment when the sun is at its highest before any place on the earth's surface, that is to say, when the sun crosses the extension into space of the "plane of the meridian" of any place, it is "local solar noon" at that place. When the sun crosses the plane of the Greenwich meridian it is at its highest in relation to Greenwich and, therefore, "solar noon" at Greenwich. Paris lies on a meridian east of Greenwich, so that it is local noon there before Greenwich and the time is said to be so much "fast" of Greenwich. Similarly, at a place west of Greenwich the local time is so much "slow" of Greenwich.

Equation of Time. As already stated, mean time is arrived at by averaging the varying lengths of the solar days throughout the year. Mean noon and solar noon at any given place agree only four times in a year; at other times, the sun is before or after the mean time clock and the numerical difference is known as the "equation of time." The maximum difference before or after the clock amounts to approximately 15 minutes mean solar time (M.S.T.). The days on or about which the clock and sun agree are April 15th, June 15th, August 31st and December 24th. These dates vary slightly owing to the fact that the year does not consist of an exact number of days. The duration of the solar year is 365 days 5 hours 48 minutes 48 seconds, so that, taking the year as 365 days involves the reckoning of an extra day once in four (leap year), and a further adjustment by missing this extra day (as in the year 1900) three times in every four centuries. The differences constituting the equation of time depend upon two factors —the position of the earth in its orbit and the obliquity of the ecliptic.

The equation of time for any day in the year is set out, amongst quantities of other valuable information about celestial bodies, in an official annual publication, *The Nautical Almanac*. This wonderful book of figures is essential to anyone wishing to make accurate observations wherewith to check the time of his clocks.

Checking a Clock by Solar Observation. A simple illustration to show how a clock may be checked by solar observation, at the same time showing the factors which have to be employed, is as follows :—

The observer must first of all possess a good sundial very carefully planted and graduated. He notes the instant the shadow arrives on the XII line (observer's local solar noon) and immediately reads the time by his mean time clock, or if the clock is not conveniently near to the sundial, his mean time watch. He then turns up the equation of time for the particular day and, if his station is situated on the Greenwich meridian, he simply applies the correction and compares the result with his clock. If, however, his station is not on the Greenwich meridian, he has also to make a correction for longitude. The sundial, in this case, is not revealing Greenwich local solar noon as well as the observer's local solar noon. Let it be assumed that observations are made from Coventry, meridian 1° 30′ W., which, converted into a time period by multiplying by 4, equals 6 minutes 0 seconds; then, when the sundial indicates XII o'clock it will have been solar noon at Greenwich 6 minutes earlier. Conversely, when it is solar noon at Coventry, Greenwich solar time is 12 hours 6 minutes 0 seconds, and if the equation of time is "sun after clock" 5 minutes, the observer's mean time clock at Coventry should indicate 5 minutes more or 12 hours 11 minutes 0 seconds Greenwich mean time (G.M.T.).

Simple though this method may appear, it is, for obvious reasons, not a very practical one. The sun has a habit of disappearing behind a cloud at the critical moment or casts a blurred shadow, so that a more reliable procedure has to be adopted.

The Transit Instrument. The most accurate observations can only be made through a telescope of a special design, known as a transit instrument (Plates I and II). This is a telescope pivoted and fixed very rigidly at an observer's station in such a way that it can only move about an east-west axis and only reveal objects as they cross the meridian and appear highest in the heavens to the observer. An adjustable frame is provided within the tube, carrying five or some other odd number of equally spaced vertical spider threads, together with one horizontal, all being situated in the focal plane of the eyepiece and object glass. One of the bearings is hollow, so that the light from a side lamp may penetrate the interior of the tube and illuminate these spider lines. The centre vertical line is adjusted in the optical axis of the instrument and an image is observed

PLATE I.—A simple form of Transit Instrument showing an oil lamp to illuminate a small reflector in the centre of the cube, giving a bright field and dark spider lines.

PLATE II.—A more elaborate Transit Instrument. Besides an oil lamp for giving a bright field, electric illumination produces a dark field and bright lines for use on faint stars. The circles, one of which indicates altitudes, whilst the other, at zero, points to the celestial equator for reading off declination direct, are also illuminated. A removable striding level and a diagonal eyepiece to facilitate observation near the zenith, can be seen.

[*To face page 10.*

to enter from the right and pass along the horizontal line to the left. The moment of transit is the time the image crosses the centre line, but the additional vertical lines can also be used as a check, the time values of their distances apart being known. If an observation of the sun is being taken, no illumination of the spider lines is necessary, but the eyepiece must be covered with a light filter, and then the moment of transit has to be ascertained by noting the time of arrival of the first limb and the departure of the second. The mean of these values gives the moment of transit of the centre.

Right Ascension and Declination. Although solar observations are useful under certain conditions, they are not sufficiently reliable for astronomers to employ for computing the accuracy of a standard clock, and to investigate these methods of time checking it will be necessary to turn once again to sidereal time.

One must first become familiar with the way in which a star or other celestial object is located in the heavens. Just as a terrestrial place, that is, a place on the earth's surface is identified by its longitude and latitude, so the direction of a celestial body is correspondingly determined by its "right ascension" (R.A.) and "declination." R.A., though an angular distance, is measured, as a general practice, not in degrees of arc but in sidereal time. As the year advances the apparent diurnal position of fixed stars varies, so that there has to be some starting point to denote the commencement of the sidereal year, just as January 1st marks the commencement of the civil year. This starting point is the vernal (or spring) equinox, when the sun appears to cross from south to north of the equator. At this moment, the sidereal clock and the mean solar clock agree, and thenceforward the sidereal clock gains 3 minutes 56·55 seconds daily over the mean solar clock.

In a sidereal year the earth completes rather more than 366 rotations (or sidereal days), which, represented in terms of mean solar time, is about 365 days 6 hours 9 minutes 11 seconds. Thus, at the end of the year, when the earth returns to its original position in relation to the stars, the vernal equinox, which occurs again on the completion of the solar year of 365 days 5 hours 48 minutes 48 seconds, will have already passed 20 minutes 23 seconds before. This curious retrograde motion of the equinoxes among the stars is known as "precession of the equinoxes," and the amount of movement, 50·25 seconds of arc, is called the "constant of precession."

The plane of a meridian imagined to be extended into space that passes through a star is called the "hour circle" of that star. In order to form the basis upon which to determine the R.A. of a star, a point in the heavens, which may be regarded as an imaginary star, is assumed to be situated exactly at the vernal equinox, that is, at the intersection of the equator and the ecliptic. This equinoctial point the ancients termed the "First Point of Aries," symbolised thus, ♈, because it was then situated in the constellation Aries. The expression is still retained, though precession has since caused it to retrograde into the constellation Pisces. The hour circle through ♈ is called the "equinoctial colure," and the R.A. of any star is its angular distance measured in sidereal time eastward between this plane and the hour circle of the star.

When the first point of Aries is on the meridian of any place, it is sidereal noon at that place and the sidereal time (S.T.) is then 0 hours 0 minutes 0 seconds. The first point of Aries is thus the zero for measuring the R.A. of a celestial body, just as the Greenwich meridian is the prime meridian for obtaining the longitude of a terrestrial place.

In order to ascertain, therefore, the moment when a star is to be seen on an observer's meridian, the position of the first point of Aries in relation to Greenwich on the particular day is required. That is also shown in *The Nautical Almanac* under the heading " Sidereal Time at Greenwich Mean Noon " ; in other words, the position of the first point of Aries at Greenwich mean noon. Now by deducting the S.T. at noon of the particular day from the R.A. of the star to be located, one obtains, in S.T., the interval which will elapse after Greenwich mean noon before the star will transit the Greenwich meridian. An observer on any other meridian must take into consideration his correction for longitude as described for solar observation. Since R.A. is measured eastward, it follows that if the value for S.T. exceeds that for R.A., 24 hours must be added before deducting S.T. to obtain this sidereal interval between Greenwich noon and the transit of the star.

Declination, the other important factor, enables the observer to discover the altitude in the heavens where he should expect to see the star transit his meridian. Declination is the angular distance, measured in degrees of arc, of a celestial body north or south of the celestial equator. This corresponds to latitude on the earth's surface.

Horizons. To understand how declination is applied to time-checking observations, there are still some other factors which require explanation. To an observer, the sky above him appears as a gigantic dome bounded by a circular horizon, he being the central figure. If he is at sea level, the horizon is described as the " sensible horizon " and a parallel plane below him passing through the centre of the earth is known as the " rational horizon." For astronomical purposes all reckonings, such as declination, are calculated from the centre of the earth, and, because of the minute size of the earth in comparison with the vast distances from celestial objects in general, the sensible and rational horizons become merged into one celestial horizon. Therefore, the rational horizon is really the basis for the calculations of an observer.

Again, if the observer is raised to a height above sea level he can see more of the earth's surface. This aspect is known as the "visible horizon."

To return to the dome, looking straight upward the point of the arch vertically above the observer's eye is called the "zenith." Any object appearing within the dome is described as at such an "altitude" from the observer, in so many degrees, minutes or seconds above the horizon. To an observer, therefore, on the earth's equator the zenith is on the celestial equator and the altitude there of a celestial body is, except for certain small corrections, also its declination. At the north pole, however, the zenith is on the axis and the horizon coincides with the celestial equator. From here nothing in the heavens is visible which has a south declination, and similarly at the south pole nothing is visible of bodies having a north declination. At all other latitudes, visibility extends partly N. and S. of the celestial equator. In general, then, an observer requires to make a permanent correction to allow for the latitude of the place when applying the declination for the purpose of finding the altitude of a star. This permanent correction is the complement of his angular distance from the equator. In other words, 90° minus his latitude. This is known as "polar distance" or co-latitude, and an observer elevating his telescope in the plane of his meridian through this angle, in effect determines the altitude of the celestial equator. The altitude at which a star crosses the meridian is simply its declination added to or subtracted from the polar distance, according to whether the declination is on the same side or the opposite side of the celestial equator.

Refraction. Right Ascension and Declination are, therefore, the two principal requirements for finding a star in the telescope. Refraction is, however, another important correction which has to be applied in making

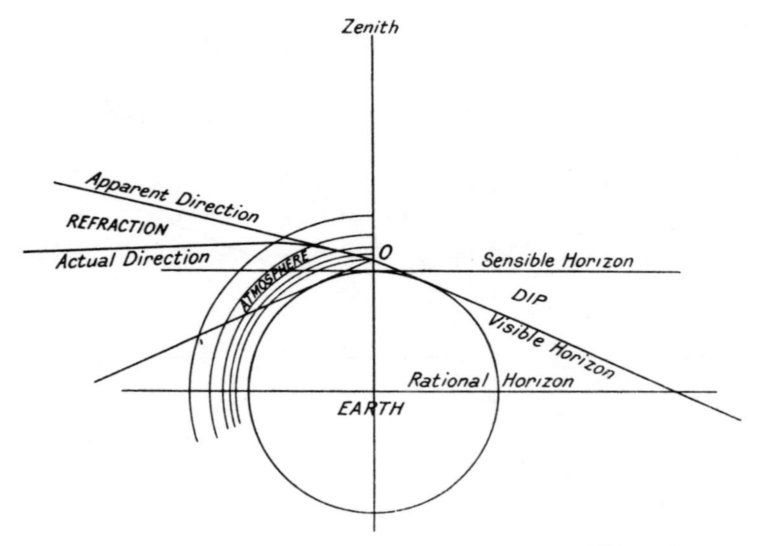

FIG. 4.—Diagrammatic representation of Horizons, Dip and Refraction.

accurate observations, apart from errors in the instrument itself, but these are eliminated by the experience of the user who makes a point of discovering and allowing for defects.

Refraction is due to the earth's atmosphere and has the effect of distorting a ray of light from a celestial object, raising it slightly towards the zenith. The error is greatest at the horizon, where the amount averages about 34′, vanishing completely when observation is towards the zenith. This correction varies with local barometric pressure and temperature and the position of the object above the horizon, and must therefore be allowed for in determining altitude.

Diagrammatic Definitions. In concluding the subject of celestial observation for time-measuring purposes, it

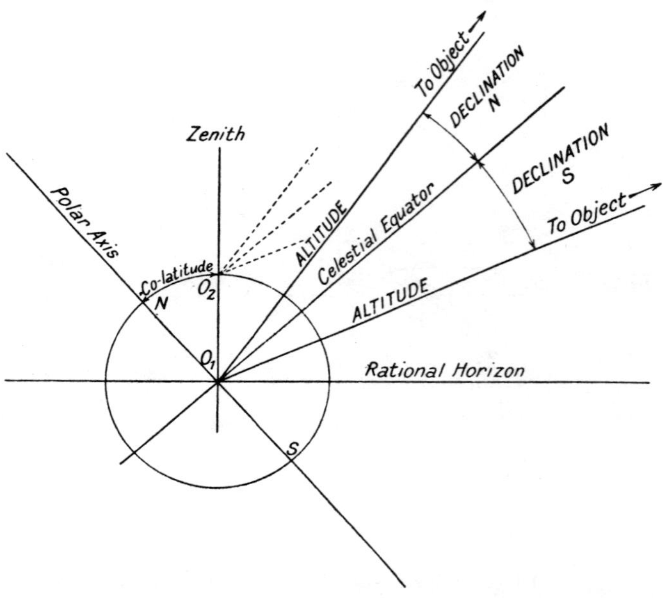

Fig. 5.—Illustrating the relationship between Declination and Altitude.

will be helpful to review the whole with the aid of diagrams, together with formulæ and a practical example. The student will easily realise how extremely difficult it is to attempt to illustrate on a small scale the immense tracts involved in a survey of the heavens, and much, therefore, depends on a realistic conception of the imaginary planes extended into space.

Figs. 4 and 5 are self-explanatory, and both are drawn

in sectional elevation. The dotted lines in Fig. 5 show the actual view to an observer, O_2, situated on the earth's surface, but, in space, these lines converge, like railway lines in the distance, into those drawn through the centre of the earth, O_1, from whence all the measurements are taken.

Fig. 6 shows the celestial sphere viewed towards the west from outside, the earth being represented by the

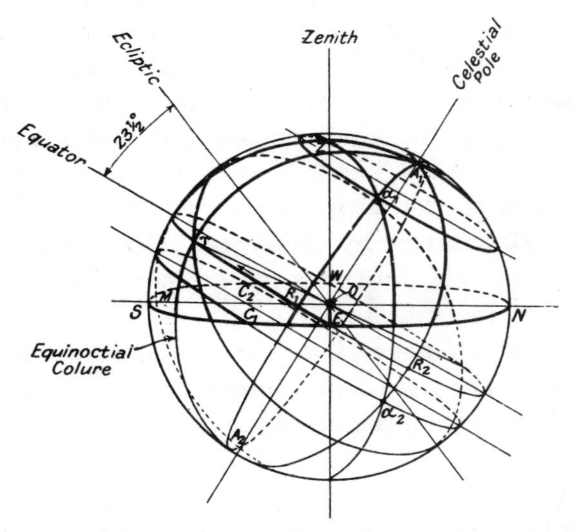

FIG. 6.—Representing the Celestial Sphere with the Earth situated at the centre and an observer, O, bounded by the Horizon *NESW*.

small circle surrounding the centre *O*. *NESW* is an observer's rational horizon, the zenith, *Z*, being shown immediately above him. No objects below *NESW* are visible to the observer. The equator and ecliptic are indicated, and the point of their intersection, ♈, is the vernal equinox or first point of Aries. At right angles to the equator is the polar axis intersecting the celestial sphere at the celestial poles A_1 and A_2, north and south respectively. Everything within the sphere appears to rotate about this axis in the direction of the arrow marked on the equator. $A_1 \Upsilon A_2$ is the equinoctial colure. α_1 and α_2 are any stars, but α_2 is below the horizon and cannot be seen by the observer until it reaches point C_1 on the rational horizon. It is there said to " rise " and it will " set " at C_2 on the other side of the sphere. The star α_1 is always above the horizon, performing a complete circle within view of the observer. A_1ZM is the observer's meridian and all objects appear to the observer highest in the heavens when they cross that line. The right ascension of α_1 is ΥR_1 and that of α_2 is ΥR_2. The declination of α_1 is $R_1\alpha_1$ and of α_2, $R_2\alpha_2$. It should be noticed that α_2 has a south declination. $A_1\alpha_1A_2$ and $A_1\alpha_2A_2$ are the hour circles passing through the respective stars.

Time Checking by Sidereal Observation. The following example illustrates the method of checking time by sidereal observation :—

An observer's latitude is 40° 30′ N. and longitude 20° E. (a) To what angle must his transit instrument be elevated for the observation of a star whose declination is 19° 36′ 50″ N.?

$$\text{Altitude} = \text{Co-lat.} + N(\text{or} - S)\ \text{dec.}$$
$$\begin{aligned}\text{Alt.} &= (90^\circ - 40^\circ\,30') + 19^\circ\,36'\,50'' \\ &= 49^\circ\,30' + 19^\circ\,36'\,50'' \\ &= 69^\circ\,6'\,50''.\end{aligned}$$

At this altitude refraction will amount to about 21″, and allowing for this, the apparent altitude at which the star transits the observer's meridian is 69° 6′ 29″.

(b) *If the sidereal time at Greenwich mean noon is 5 hours 21 minutes 54 seconds and the right ascension of the star is 14 hours 11 minutes 50 seconds, at what G.M.T. will the transit occur on the observer's meridian?*

	H.	M.	S.	H.	M.	S.	
R.A. of star (that is, S.T. of transit of star) .				14	11	50	
S.T. at Greenwich (at Greenwich mean noon) . . .	5	21	54				
20° E.	1	20	0	6	41	54	
S.T. at place of observation at Greenwich mean noon				7	29	56	S.T.

S.T.	G.M.T.						
7 Hrs. =	6	58	51·19				
29 Mins. =		28	55·25				
56 Secs. =			55·85	7	28	42·3	G.M.T.

CHAPTER II

TIME RECORDING

Constituent Parts of the Mechanism. From the fundamental considerations of time and its measurement, it now becomes necessary to pass to the mechanical features of instruments used in its automatic record. The essential constituents of such a mechanism are, first, some form of motive power, and, secondly, some system of controlling and regulating the expenditure of that power into a uniform flow at a definite speed. In a sand glass, the attraction due to gravity is the motive power and the small hole in the neck of the tube is the controlling influence. In modern clocks and similar contrivances, the motion is restrained by the intermittent action of the part known as the escapement, and controlled by the swing of a pendulum or the vibration of a balance and its spring. The motive power is generated either by means of a raised weight or coiled spring, or by an electrical device. Subsidiary to these two principal features there is the train, which consists of a series of wheels and pinions communicating the energy to the escapement and to the time-measuring property, and the motion and dial work for producing the visual time record. In some cases the latter may operate in conjunction with a striking mechanism and, according to requirements, may even take the form of an electric signal, a paper reel or some other desired method of recording.

Historically, contrivances for recording time have passed

naturally through stages of evolution, but the greatest technical advancement did not appear until about the seventeenth century. Of the ancient devices, Clepsydræ, or water clocks, were perhaps the most important. In these, intervals of time are noted by the trickle of water from a bowl or similar receptacle into a graduated vessel below. They were in use in Egypt, Babylonia, China, India and other eastern countries several centuries B.C. In some improved forms, as shown in Fig. 7, the lower vessel was cylindrical in shape, and as the water collected it raised a float which, by means of a rack and pinion, operated a single hand round a dial. Sundry other improvements were added from time to time even up to the seventeenth century. One of these later examples can be seen in the Victoria and Albert Science Museum at South Kensington.

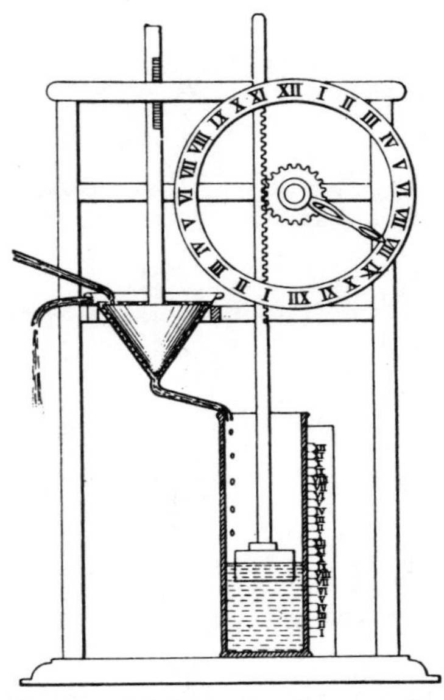

FIG. 7.—A "Clepsydra"; an ancient device for denoting Time by regulating a trickle of water.

Lamps with a graduated oil reservoir, as well as sand glasses, were also early devices for denoting the passage of time.

The date of the actual introduction of clocks consisting of a train of gear wheels driven by a weight is most uncertain, especially as the term was obviously often used to describe a sundial. Relics of early types still remain, which show that they were controlled by a horizontal balance, or crossbar, called a "foliot." As shown by

Fig. 8, this was merely a straight arm colleted at the centre to an upright arbor, which was suspended from a cord at the top and pivoted at the foot into a block projecting from the frame. Two flat pallets were wrought into this arbor, or "verge," as it is technically known, and these engaged with the ratchet-shaped teeth of an escape wheel mounted on an arbor at right angles to the verge. An adjustment for closer timing of the vibration of the balance arm was provided by two adjustable weights suspended from notches near each end.

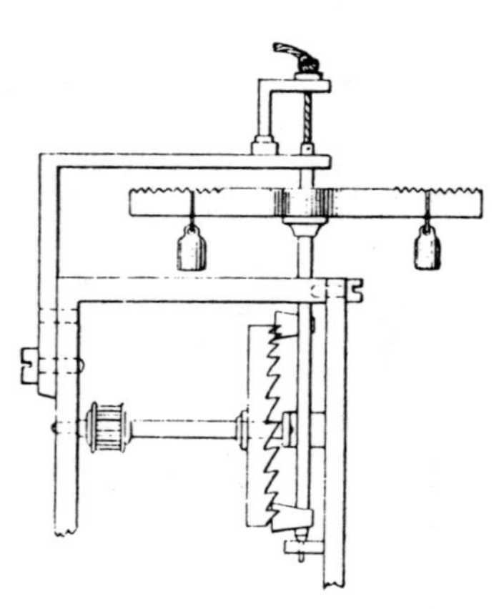

FIG. 8.—A "foliot"; the earliest form of mechanical clock escapement.

No tangible records appear to go farther back than the thirteenth or fourteenth centuries, when some very crude mechanisms on this principle were designed. One of the earliest examples was thought out by a Glastonbury monk, Peter Lightfoot, about the beginning of the fourteenth century, for the Abbey. The movement, though altered from its original form, is now preserved in South Kensington Museum. At the time of the dissolution of the monasteries the clock was removed to Wells Cathedral, and it may have been then refitted with the escapement now to be seen at the Museum, but the dial, curiously befigured and also having undergone restoration at different times, is still in use on the present modern clock in Wells Cathedral.

Another important figure among the records of early horologists was Heinrich von Wieck, who, about the year 1364, received an order from Charles V of France to construct a turret clock for his palace in Paris. This clock was similar to the Glastonbury clock and still exists in the Palais de Justice, though having been restored and undergone changes from time to time.

Clocks appear to have been designed on this principle for about three hundred years.

In the year 1581, the renowned Italian physicist Galileo Galilei (1564–1642) entered the University of Pisa, where he remained until 1585. During that time he made the important discovery of isochronism in the swing of a pendulum, which is referred to more fully in Chapter III, on Pendulums. It was not, however, until the year before his death (1642) that the pendulum was actually applied by him for the measurement of time. He designed an escapement which was subsequently constructed by his son, Vincenzio Galilei, to demonstrate the worth of the pendulum, though it was ultimately to the verge escapement that its application became general. Thus it remained until long after Galilei's time, whilst the verge escapement and circular balance were not superseded, in portable clocks and watches, before the middle of the eighteenth century.

Many inventions were made during the eighteenth century, and the names of Sully, Harrison, Graham, Mudge, Arnold, Earnshaw, Julien and Pierre Le Roy, Berthoud, Bréguet and others became eminently associated with horology. Often the ideas of one man, not being perhaps fully appreciated by himself, were applied or brought to perfection by another, and, in some cases, similar discoveries seemed to spring up concurrently among different inventors. Contributory to accurate navigation, the necessity of time-keeping instruments of the utmost precision initiated considerable scope for horological inventive genius, and it is a fact that most of the constructive principles founded during that period have not been superseded by more modern science. Instruments have, of course, been embellished by subsequent methods of production and by the introduction of machinery for developing the principles with greater accuracy. Details of the precise nature of various discoveries will be given later in the appropriate sections to which the innovations refer. The general scope of the sections is as follows :—

Clocks. Usually time-recording instruments in which the intervals are controlled by a pendulum. Also used to denote mechanisms of a smaller size controlled by a balance and balance-spring, employed either alone or in conjunction with some other instrument.

Watches. Small portable instruments for personal wear or adornment and various other purposes.

Chronometers. Instruments of great precision, used chiefly as navigational equipment, provided with the detent-chronometer escapement.

PART II—CLOCKS

CHAPTER III

PENDULUMS

Galilei's Discovery. As already stated, it was the middle of the seventeenth century when the pendulum was first applied to time-recording instruments. Between the years 1581 and 1585, Galilei, whilst in Pisa Cathedral, observed by comparison with the beating of his pulse that the swings or vibrations performed by the large hanging lamp, whether long or short, were of uniform duration. This discovery led to the derivation of the term "isochronism," which means "equal time." In the Science Museum at South Kensington there is a model of an escapement constructed on the lines of Galilei's original idea to show how a pendulum could be applied as a counter of time intervals. This particular method was never, however, brought into general use, though it resembles closely the "pin-wheel" escapement, a description of which appears in Chapter IV, on Clock Escapements.

Huyghen's Demonstration. About the year 1665, the distinguished Dutch mathematician, Christian Huyghens (1629–1695), demonstrated that the arcs described by a pendulum were not truly isochronal unless the path was that of a cycloid instead of a circle. A pendulum describing very long arcs, as those used in conjunction with the verge escapement of that period very naturally would do, was found to take longer than when describing only short arcs. Huyghens endeavoured to surmount this difficulty by introducing curved "cheeks" on either side of the suspension, thus forcing the pendulum to describe the path of a cycloid as it vibrated. This experiment, however, produced considerable friction and did not find favour, because it soon became far easier to confine the vibrations within much narrower limits.

A pendulum is a body which is free to oscillate about an axis that does not pass through its centre of gravity, and of such there are three classes:—(1) Simple: (2) Compound: and (3) Compensated.

Simple Pendulums. A "simple" pendulum is the theoretically perfect one; namely, a single heavy particle of mass suspended by a flexible weightless cord. A heavy bullet hanging by a fine thread is perhaps the nearest example of this. When set in motion it continues to swing for a very long period, the force of gravity supplying it with sufficient energy during its downward motion to enable it to rise a corresponding amount when travelling upward. This form of pendulum is the idealistic and mathematical conception, and not of practical value.

Compound Pendulums. A "compound" pendulum is composed of many particles and appertains to all those used in clocks. They generally consist of a thin steel suspension and a rod with a heavy bob at the bottom and the tendencies of the different particles, each endeavouring to vibrate in its own time along its own arc, those nearer the top faster and those nearer the bottom slower, are resolved into one at a point called the centre of oscillation. This point is situated in the bob and its distance from the centre of suspension is identical with the length of the "simple" pendulum which would give the same periodic vibration.

Compound pendulums suffer certain disturbing influences which must be carefully taken into consideration. In the first place, there are two causes which will bring a swinging pendulum to rest :—

(1) Resistance offered by the density of the air, and
(2) Friction at the suspension.

Without these a pendulum would remain in perpetual motion.

Then, in the case of a clock pendulum, other troubles arise which tend to cause variation in the time-keeping, namely:—

(1) Changes of temperature (" temperature error "),
(2) Difference in the length of arc described (" circular error "), and
(3) Interference of the escapement.

Barometric error, being small, is not a serious matter in the case of ordinary clocks, though for very accurate astronomical regulators it does become an important consideration. The usual method of reducing this error to a minimum is to mount the clock in an airtight case, by means of which a very low pressure, almost a vacuum, can be maintained."

Friction at the suspension is reduced as far as possible by using a short thin piece of ribbon steel. The residual friction is then due to the internal displacement of the molecules in the process of bending, known as molecular friction. It should be mentioned that isochronal adjustment is made possible by cutting a hole of appropriate dimensions in the steel suspension spring. Other forms of suspension are sometimes employed, such as silk or knife-edge, but they generally have disadvantages where clockwork is concerned.

Circular error, again, is not great and is reduced to a minimum by making the pendulum describe the shortest arcs possible. As previously mentioned, when Huyghens discovered the error only the wide-swinging pendulums necessitated by the verge escapement were in general use. Now the average arc of a seconds pendulum used with a dead-beat escapement is confined to $2\frac{1}{2}°$ on either side of the vertical.

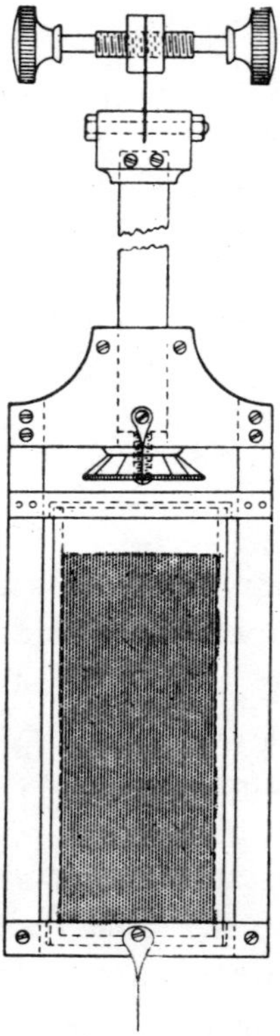

Of all disturbing influences, by far the most serious is "temperature error." The effect of changes in temperature is to lengthen or shorten the distance between the centre of oscillation and the centre of suspension, which is the sole feature governing the time of vibration. Methods of overcoming this difficulty are embodied in the third class of pendulum—"compensated pendulums."

Fig. 9.—A Mercury Compensated Pendulum based on the principle originated by George Graham.

Compensated Pendulums. Compensation is effected by using for the composition of the pendulum a combination of materials having different coefficients of expansion. Changes of temperature cause the materials to react with each other, and the effective length of the pendulum is thus maintained as nearly constant as possible.

Perhaps the most important of the early inventions was made by George Graham (1673–1751) at the beginning of the eighteenth century. A glass jar partly filled with mercury forms a bob, the rod is of steel and carries the jar in a stirrup, as shown in Fig. 9. The upward expansion of the mercury tends to counteract the effect of the downward expansion of the rod. This form of compensated pendulum has found considerable favour, though generally the stirrup arrangement is modified by attaching a steel jar to the rod.

Fig. 10 illustrates another important pendulum, invented by John Harrison (1693–1776) about the same time, and known as the "gridiron." This consists of five steel and four brass rods arranged alternately and coupled together in the manner shown. The downward expansion of the steel counteracts the upward expansion of the brass. The "gridiron" pendulum came into more extensive use on the Continent than in this country.

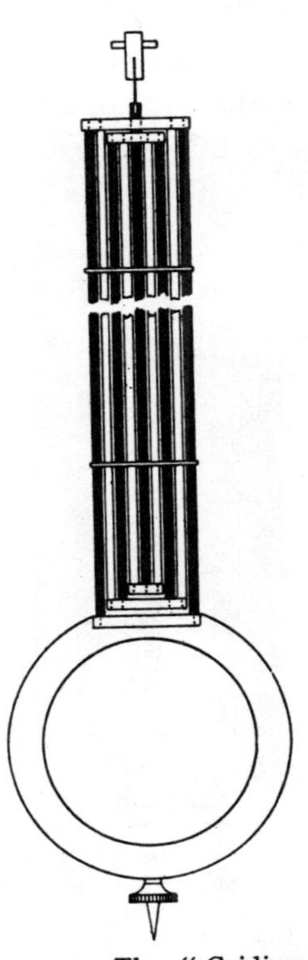

FIG. 10.—The "Gridiron" Pendulum; Compensation being effected by alternate steel and brass rods coupled together. The invention of John Harrison.

A third type is a combination of zinc and steel tubes. The rating nut at the bottom of a steel rod carries a zinc tube which supports from the top an outer tube of drawn steel. At the lower end of this, a brass collar forms a support for a lead bob bored so as to be suspended from the centre. To admit a free passage of air, the steel tube is slotted and the zinc tube drilled at intervals. This is considered to be the easiest form of compensation.

The more recent discovery of the nickel-steel alloy, "Invar," by Dr. C. E. Guillaume has, however, to a considerable extent revolutionised compensated pendulums. This curious metal is hardly affected by change of temperature at all—indeed, some grades have been known to possess a negative coefficient of expansion, *i.e.*, to expand in cold and contract in heat.

A very effective way of constructing the pendulum is shown in Fig. 11. A cylindrical steel bob drilled lengthwise to admit the invar rod is bored wider from the bottom to the centre so as to provide a seating for a short steel pipe forming the upper part of the rating nut. The foot of the rod is tapped and received at the top of the pipe. The changes in the bob are thus, as in the zinc and steel pendulum, independent of the behaviour of the rod, and the short pipe and extra washers, if necessary, form a means .of compensating the effect of temperature on the suspension spring.

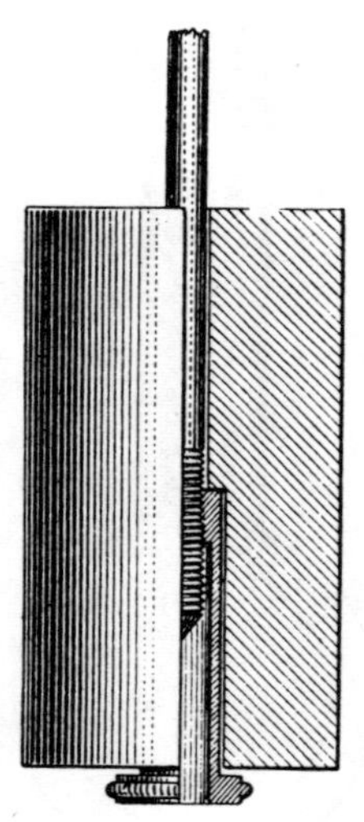

FIG. 11.—"Invar" Compensation; showing the method of supporting the bob.

Free Pendulums. The demand for the greatest possible accuracy in astronomical work has given rise to considerable scientific effort in the endeavour to obtain a pendulum perfectly free from external disturbing influences, particularly that of interference of the escapement.

In 1891, Dr. Siegmund Riefler, of Munich, produced a clock with a free escapement and a mercurial pendulum with a mannesmann-s⁺eel rod, which has achieved very remarkable success in this respect.

The impulse is delivered to the pendulum through the suspension spring and consequently the usual crutch acting on the rod is absent. Fig. 12 shows a front and side view of the method of supporting the pendulum. Two escape wheels, one, E_1, with radial acting faces to the teeth for locking and the other, E, with sloping acting faces for

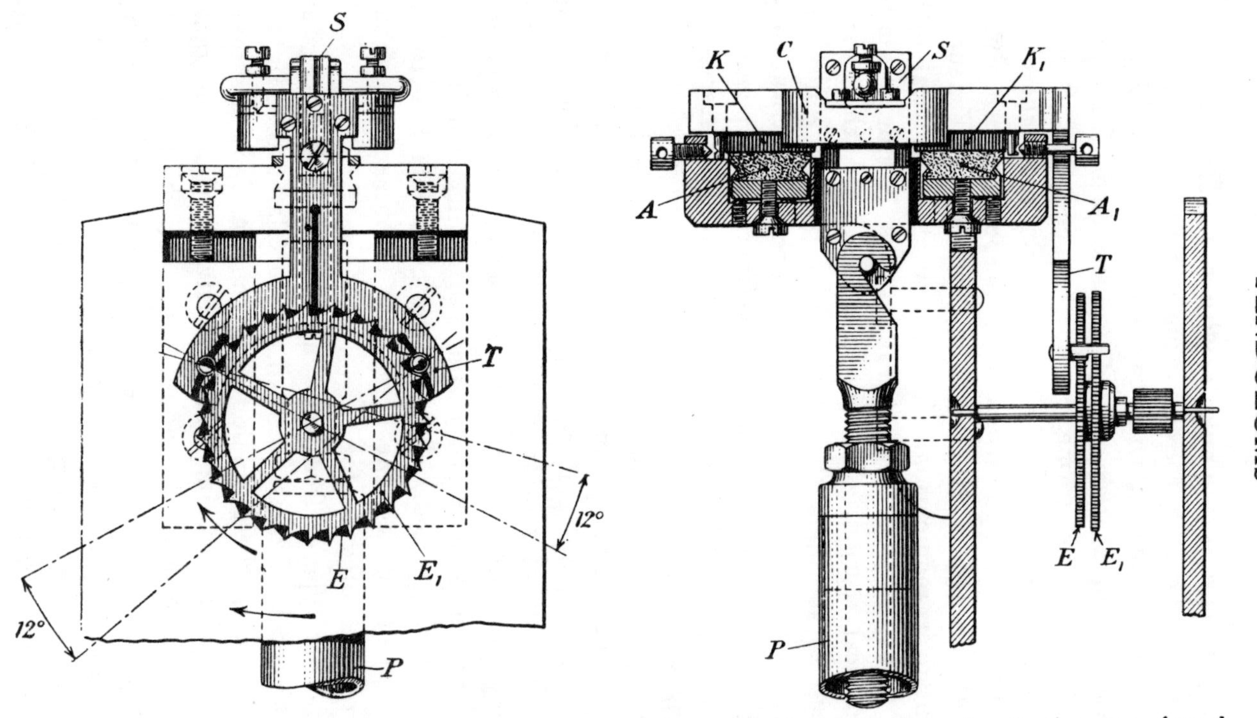

FIG. 12.—The "Riefler" Free Pendulum. Impulse being delivered through the Suspension Spring the vibrations are performed unhampered by the friction of a crutch.

giving impulse, communicate with the pallets, *T*, which carry a short crutch, the end being attached to a horizontal cradle, *C*. Provided with two knife edges, *K*, K_1, this cradle rides on agate pieces, *A*, A_1, and forms the support for the suspension spring, *S*, and the pendulum, *P*. The sideways movement of the pallet crutch rocks the cradle and produces a slight strain in the suspension spring with every vibration of the pendulum, which is sufficient to maintain its momentum. The pallets are cylindrical with the front half cut diametrically across leaving a semi-cylindrical section. The impulse teeth of the escape wheel, *E*, act upon the rear circular portion whilst the radial teeth of the wheel, E_1, lock on the diametrical faces in front. As shown in the front elevation the pendulum is swinging to the left, and the crutch, cradle and suspension spring will proceed together until the tooth of the escape wheel, E_1, has dropped off the left-hand locking. The right-hand impulse pallet then receives the next tooth of the wheel, *E*, which forces the crutch and cradle against the direction of motion of the pendulum until the next locking tooth drops on the locking face of the same pallet. The locking is released and impulse commenced in both cases, almost immediately after the pendulum has crossed the line of centres, the actual arc amounting to about one-quarter of a degree.

Another type of free pendulum, shown in Fig. 13, was introduced by Professor L. Strasser, of Glashütte. In this case, however, there is a double suspension, *S*, S_1, the lower portions of each being clamped between the same brass chops upon which the pendulum is hung, whilst the inner-most upper chop carries the pivots for mounting on the suspension block, and the outer chop, consisting of two sides and a connecting member above, is free. In the middle of the top member an agate centre is fixed, into which projects a pivot from the carriage, *C*. Actuated thus by the sideways movement of the carriage, *C*, and the pallet crutch, *T*, this free or impulse spring produces an impelling effect which is enough to meet the requirements of the pendulum. The locking occurs at the end of the beat, and then follows the reverse action.

The accuracy of performance of Riefler clocks, for many years, has proved superior to that of any other for astronomical purposes and only recently have these clocks come

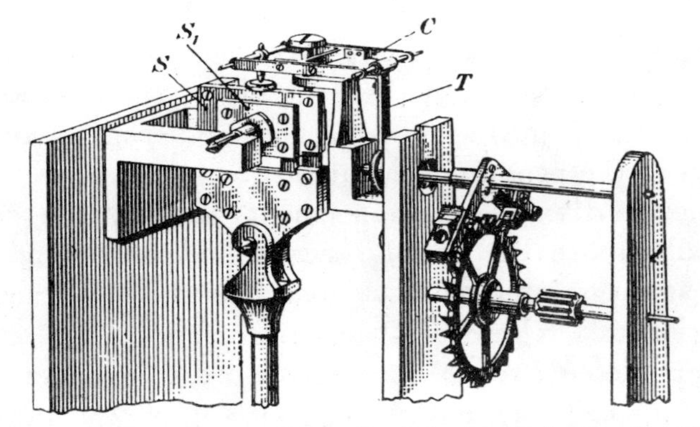

FIG. 13.—The "Strasser" Free Pendulum; showing the Double Suspension which is used for transmitting impulse.

into competition with the "Shortt," which is described later in this chapter. As a self-contained clock, independent of electric currents, the Riefler has only been equalled by the Strasser, which presents the simplification of a single escape wheel.

It is a fact of interest worth noting that Dr. Riefler, in 1898, employed nickel-steel for the first time in the construction of pendulums.

Various other forms of double-suspension free pendulums of German origin have also been introduced, but actuated electrically instead of by means of an escapement.

To mention two of these, one was patented by Edmund

Pfeiffer, of Dresden, and another by Ed. Schlesser. Both have been described in the *Deutsche Uhrmacher-Zeitung*.

In Pfeiffer's system, the top suspension chop carried a short crutch vertically above the rod, the end of which passed between two opposite contact points. These were the tips of screws projecting from angle pieces mounted below a horizontal slide, free to move equally a short distance on either side of the rod. The sideways movement of the slide was produced by the attraction of an armature, mounted above, to two independent coils through the "make and break" contact at the end of the crutch. Thus, first one way and then the other, the impulse suspension was strained against the normal suspension and the motion of the pendulum maintained.

Schlesser devised an arrangement whereby the crutch vertically above the impulse suspension chop terminated in an armature and was oscillated a fixed distance by the attraction and release of an electro-magnet. The circuit was kept closed and no "make and break" contacts were used, but half-way down the rod a lens was provided which, as the pendulum passed the middle position, focussed a ray of light upon a selenium cell situated at the back of the case. The momentary increase of light, natural or artificial, on the selenium, which has the peculiar property of varying its conductivity as the square root of the intensity of light upon the object, increased the flow of current through the electro-magnet and the armature was attracted accordingly and impulse given to the pendulum.

A most important departure from the usual method of designing free pendulums was introduced by C. O. Bartrum in 1917. He demonstrated the good time-keeping that would result from using two pendulums, one as a "master" and the other as a "slave," the slave to relieve the master of all work and yet remain under its control.* Impulse was supplied to the master through the descent of a roller, mounted on the end of a gravity lever, upon the inclined face of a pallet attached at the foot of the master pendulum, as shown in Fig. 14. This gravity lever was

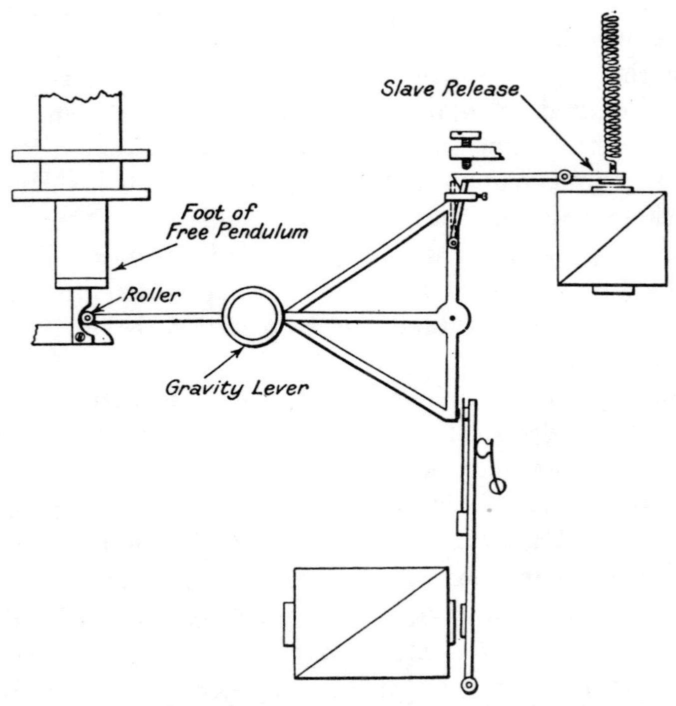

FIG. 14.—The "Bartrum" Master and Slave system; showing the Gravity Roller Impulse device at the foot of the Master Free Pendulum.

released electromagnetically once a minute by the slave. After giving the impulse, the tail of the gravity lever dropped on to a contact which closed a circuit and was restored thereby to its original position. At the same time, this contact was also used to reveal any error, positive or

* *Proceedings of the Physical Society of London*, Vol. XXIX, Part II, February 15th, 1917.

negative, in the synchronous motion of the slave in relation to the master. Should the two pendulums not be in unison, the rate of the slave was altered by the automatic increase or decrease in the tension of a helical spring connected with the slave. This method of synchronising the two pendulums did not alone yield perfect unison. The incessant tendency of the slave to be affected by external influences and deviate from the master, followed by the automatic correction of error, resulted in a continuous chasing of error and rate. This condition, known as "hunting," was prevented by the use of a supplementary rating spring which gradually damped out these periodic variations of error and rate. The master pendulum was designed to be suspended in an air-tight case confined in a chamber maintained at constant temperature, whilst the slave was intended to be placed in the observatory or wherever required.

Although quite independent attempts to carry out the principle of electrically impelling a free pendulum at the dictation of another, which the free pendulum maintains synchronous, had been made both by the late Sir David Gill, of the Cape Observatory, and R. J. Rudd, the clock devised by Bartrum was the direct forerunner of that devised by W. H. Shortt, which has achieved merit hitherto unparalleled. Shortt improved upon Bartrum's design in two features, first, by a simpler method of synchronising the slave with the master, and, secondly, by reducing the air pressure in the case of the master pendulum.

The "free-pendulum" in this combination receives impulse every half-minute by the electromagnetic release of a gravity roller allowed to descend an inclined plane attached to the rod.* This switch impulse mechanism, as well as the "slave clock" itself, is the system developed by

* *Journal of the Royal Society of Arts*, May 23rd, 1924, "The Free Pendulum," by F. Hope-Jones.

the Synchronome Co., and described fully in Chapter X, although its application to the free-pendulum is slightly modified because the impulse necessary is extremely small. A diagram of the circuit is shown in Fig. 15, but the feature of great importance is the synchronising device between

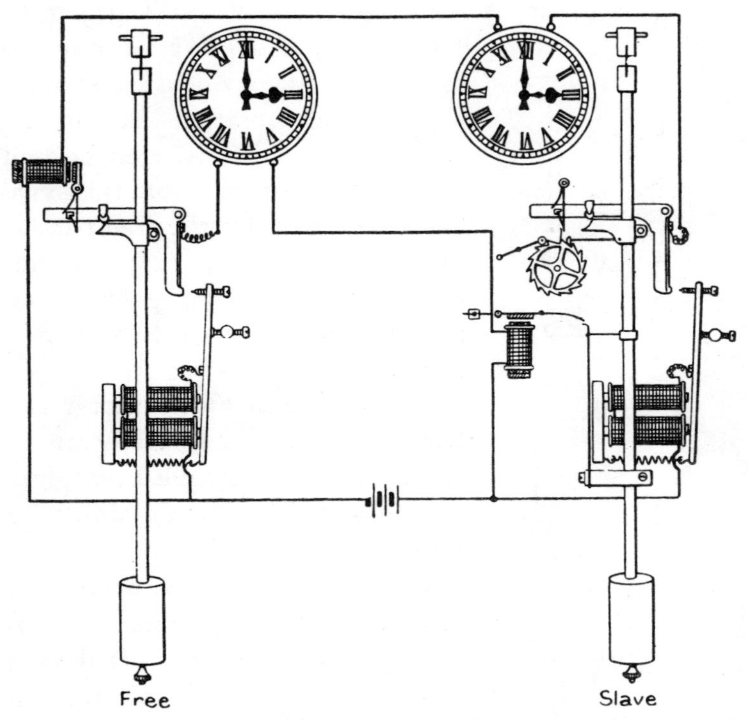

FIG. 15.—The " Shortt " Master and Slave circuit. In practice the gravity impulse lever of the Free Pendulum is very light and actuates a trigger which releases an independent resetting lever.

the two pendulums. In Fig. 16 an illustration is given of the underlying principle. The diagram shows an electro-magnet, *M*, and a horizontal armature, *A*, the position of which is adjustable by means of the set-screw, *Y*. The slave clock pendulum, *D*, carries a long vertical spring lever, *L*, normally supported by the stop *X*. The gravity lever of the free-pendulum (Fig. 15), after giving impulse, makes a contact which closes the circuit of *M*. The armature, *A*, is thus attracted momentarily into the path of the spring lever, *L*, and dependent upon the position of the pendulum the spring lever, *L*, either hits or misses the armature in passing the middle position. If engagement takes place, the spring lever acts as an aid to gravity and accelerates slightly the vibration, but if the swing has already advanced beyond the middle position before the signal has been given by the free pendulum, then the armature will not engage with the spring lever and no assistance is given to the slave pendulum.

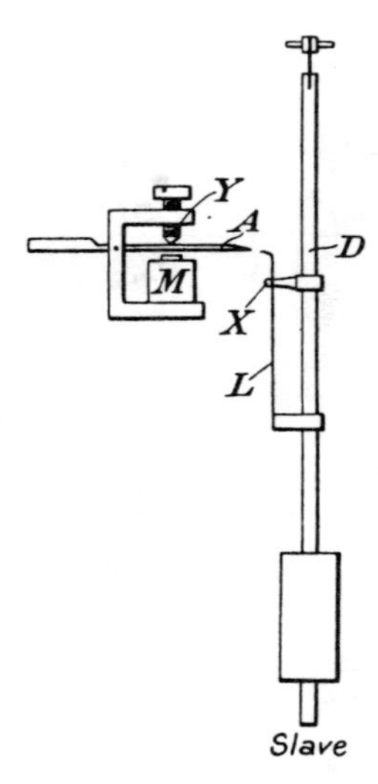

FIG. 16.—Diagram showing the principle contained in the "Shortt" synchronising device between Master and Slave.

At Edinburgh Observatory, and later at Greenwich Observatory, clocks on this principle have given amazing results, maintaining a steady rate over long periods and raising to a very marked degree the accuracy with which time measurement can be performed. It should, however, be remembered that in order to ensure good performance all clocks used in astronomical work receive continuous attention and nursing.

Theory and Formulæ of Pendulums. The usual method for mathematically determining the length of a pendulum is based on the formula :—

$$T = 2\pi\sqrt{\frac{l}{g}}$$

where, T, is the periodic time of one complete (or double) vibration in seconds, l is the length, in feet, from the centre of suspension to the centre of oscillation, and g the acceleration due to gravity, in feet per sec. per sec.

A simple way of understanding the derivation of this formula is through the laws of falling bodies and of simple harmonic motion.

In Fig. 17, if P is a point on the circumference of a circle around which it is assumed to be moving with uniform

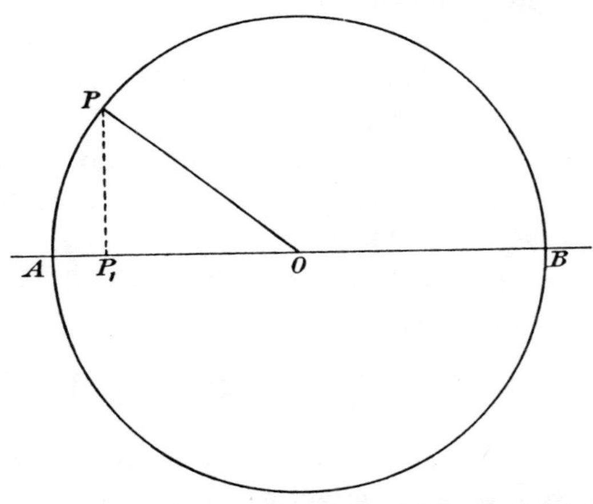

FIG. 17.—Illustrating the law of Simple Harmonic Motion.

velocity, then its motion projected on to a diameter AB is apparently backwards and forwards, and is "simple harmonic." An example to make this clearer may be realised by looking along the plane of a wheel revolving uniformly at a pin projecting above the periphery. The pin appears to pass to and fro in a straight line, slowest at the extremities of its movement and fastest in the middle. The apparent motion of the pin is simple harmonic motion.

Remembering the dynamical formula: Velocity = Space/Time, let us return to Fig. 17, where P is imagined to be travelling round the circumference at a uniform rate, once in t seconds, then :—

$$\text{Velocity of } P = \frac{2\pi OA}{t} \quad . \quad . \quad . \quad . \quad (1)$$

Now the angular velocity of OP, in radians * per second,

$$= \frac{\text{Velocity of } P}{OA} \quad . \quad . \quad . \quad . \quad . \quad (2)$$

So that, by combining the two equations, we have,

$$\text{Angular velocity of } OP = \frac{2\pi}{t}. \quad . \quad . \quad (3)$$

If P now be regarded as a particle of mass (m) at P_1, travelling backwards and forwards along AB, then, by the laws of harmonic motion, it can be shown that the acceleration of P_1 along AB = (angular velocity of $OP)^2 \times$ displacement of P_1 from O. So that, by substituting the value of the angular velocity of OP obtained in eqn. (3) above, we have—

$$\text{Acceleration } (A) = \left(\frac{2\pi}{t}\right)^2 \times \text{displacement } P_1O,$$

and the period of vibration (t) of P_1, along AB, *i.e.*, from A to B and back to A, is given by—

$$t = 2\pi\sqrt{\frac{\text{displacement } P_1O}{\text{acceleration}}} \quad . \quad . \quad . \quad (4)$$

In the case of a simple pendulum where the arcs described are small, it may be assumed that they are chords of a circle whose centre is at the centre of suspension.

* A *radian* is the unit of circular measurement of angles and corresponds to the angle whose arc is equal in length to the radius. In other words, the angle subtended by a certain arc expressed in radians $= \frac{\text{length of the arc}}{\text{radius}}$, so that, if the arc be a complete circle, then the angle measures $\frac{2\pi r}{r} =$ 2π radians. Supposing a wheel makes 10 revs. per sec., since 1 rev. = 2π radians, then the *angular velocity* of the wheel = 20π radians per sec.

In Fig. 18, let O be the centre of suspension, or the centre of the circle about which the mass P is free to vibrate. Then the acceleration of P in the direction PQ is the acceleration of gravity resolved in this direction, that is—

$$\text{Acceleration} = g\frac{d}{l}.$$

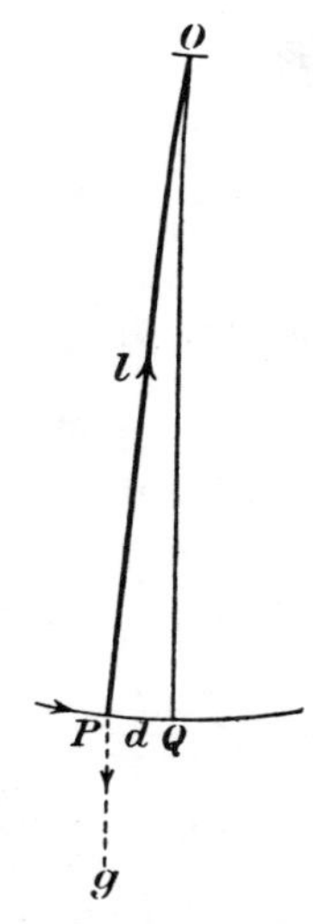

FIG. 18.—Illustrating the motion of a Pendulum.

Since here the acceleration is proportional to the displacement, the motion is simple harmonic and the time of vibration (T) by eqn. (4) is—

$$T = 2\pi\sqrt{\frac{\text{displacement}}{\text{acceleration}}}.$$

Therefore, by substitution—

$$T = 2\pi\sqrt{\frac{d}{(gd/l)}}$$

$$= 2\pi\sqrt{\frac{l}{g}},$$

and the time for a single beat ($\frac{1}{2}T$) is $\pi\sqrt{\frac{l}{g}}$. It should be noted that the value of g varies according to the latitude and elevation above sea level of the place concerned. In England it is taken as 32·2 feet per second per second. The length l of the pendulum must then be expressed in feet.

EXAMPLES.—(1) *Required the length of a seconds pendulum.*

From the formula $T = \pi\sqrt{\frac{l}{g}}$

$$1 = 3{\cdot}1416\sqrt{\frac{l}{32{\cdot}2}}$$

$$1^2 = \frac{(3{\cdot}1416)^2 l}{32{\cdot}2}$$

$$\therefore l = \frac{32{\cdot}2}{(3{\cdot}1416)^2} \text{ ft.}$$

$$= \frac{32{\cdot}2}{9{\cdot}8696} \text{ ft.}$$

$$= 3{\cdot}262 \text{ ft. or } 39{\cdot}14 \text{ ins.}$$

The working out of an example to find an unknown length of a pendulum can be simplified in some cases by noting that in the formula—

$$T = \pi\sqrt{\frac{l}{g}},$$

π and g are both constants, so that—

$$T \propto \sqrt{l}$$

or

$$T^2 \propto l.$$

Therefore, knowing the length of a seconds pendulum, the length of any other pendulum can be determined simply by ratio.

(2) *Given the length of a seconds pendulum as* 39·14 *inches, what is the length of a half-seconds pendulum?*

$$1^2 : 39{\cdot}14 :: (\tfrac{1}{2})^2 : l$$

$$\frac{1}{39{\cdot}14} = \frac{(\frac{1}{2})^2}{l}$$

$$\therefore l = \frac{39{\cdot}14}{4}$$

$$= 9{\cdot}78 \text{ ins.}$$

This formula, of course, involves a knowledge of the value of one or other of the factors T and l. It may happen in practice, however, that both are unknown. Under such circumstances, it is necessary first to ascertain the value of T, the time of vibration, from the ratio of the train wheels and pinions. This procedure is described in Chapter V.

CHAPTER IV

CLOCK ESCAPEMENTS

THE escapement of a clock is the contrivance for controlling the expenditure of motive energy and for maintaining the motion of the actual time-counter, the pendulum.

In all escapements, whatever the particular method of construction, there are two distinct operations performed by the pallets, which act as buffers to the progress of the escape wheel. The first operation, known as "locking," is actually to stop the train; the second, known as "impulse," is to give the necessary impetus to the pendulum. In most cases there are two pallets, each of which is so formed as to provide for both locking and impulse.

The Verge Escapement. The earliest form of escapement, the "verge" or "crown wheel," has been referred to already and, being now obsolete, is hardly worthy of a detailed description. It is, however, still met with in genuine antique bracket clocks and the like, and Fig. 19 will serve to show the action of the escapement.

The escape wheel is mounted vertically between the frames, and the pallets projecting from a pivoted arbor or verge have flat surfaces which engage with diametrically opposite teeth. The angle subtended between the pallet surfaces generally amounts to about 90° or 100°, but there seems to have been considerable diversity of opinion as to the most desirable proportions for this escapement.

The arrow-heads on the figure indicate the direction of motion of the wheel, pallets and pendulum, and it will be observed that the latter is compelled to make very wide arcs of vibration. For the escapement to be in beat, the pendulum obviously occupies a position midway between the angle subtended by the pallets. When a tooth has escaped, the diametrically opposite tooth drops on the other pallet, but the pendulum, having up to that point received impulse, continues its motion a little farther and in

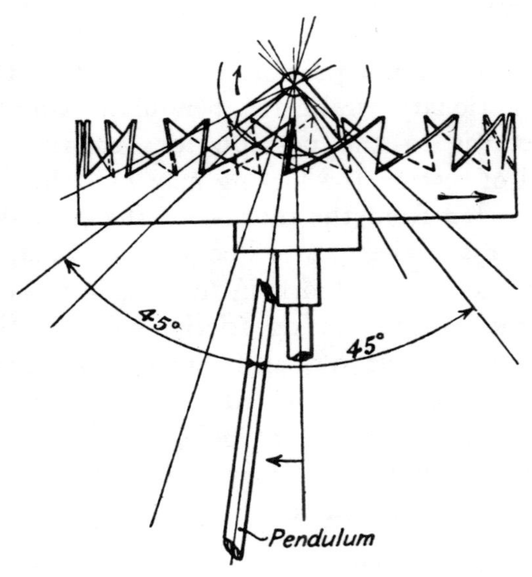

FIG. 19.—The verge or crown wheel clock escapement.

so doing drives slightly backwards the tooth which has just dropped. This constitutes the feature described as recoil. This escapement has the great disadvantage of requiring considerable force to give the necessary impulse, consequently the pallets are subjected to very severe wear and soon become badly pitted. Another unfortunate feature is that this escapement is extremely sensitive to fluctuations of motive power, which seriously affect consistent timekeeping.

Classification of Escapements. Primarily, escapements fall into two classes, namely, " recoil " and " dead-beat," but the latter are subdivided into " frictional rest " and " detached " types.

In escapements which recoil, the wheel is driven backward slightly, as in the verge, by the continued motion of the pendulum after impulse has been given and the next tooth has fallen on the locking. Dead-beat escapements show no such recoil, the locking faces of the pallets and the teeth of the wheel being so formed as to cause the wheel to remain stationary whilst the pendulum completes its arc and returns to the position for receiving impulse. This period of rest on the locking may be either *frictional* or quite *detached* from the motion of the pendulum. In the former case, the power behind an escape wheel continues to press each tooth in turn against the moving surface of the pallets, causing friction—but in the latter case the powei is cut off, as it were, from the pallets whilst the pendulum is permitted to swing freely.

Clock escapements only are being dealt with in this chapter, and their names, grouped into their respective categories, are given in Table I.

TABLE I.—CLOCK ESCAPEMENTS

RECOIL	DEAD-BEAT	
	Frictional rest	Detached
Verge, or Crown wheel. Recoil or Anchor. Half-dead. Pin Pallet (with radial or forward-cut teeth to escape wheel).	' Graham ' dead-beat. Pin wheel. Pin pallet (with backward-cut teeth to escape wheel).	Gravity.

The Recoil Escapement. The recoil, or anchor escapement, as it was originally called, is shown in Fig. 20. This was invented by the eminent scientist, Dr. Hooke (1635–1703), about the year 1656 and apparently almost immediately after the introduction of the pendulum. From that time to the present it has remained the most used form of escapement in domestic clocks and there has been very little deviation from the original design.

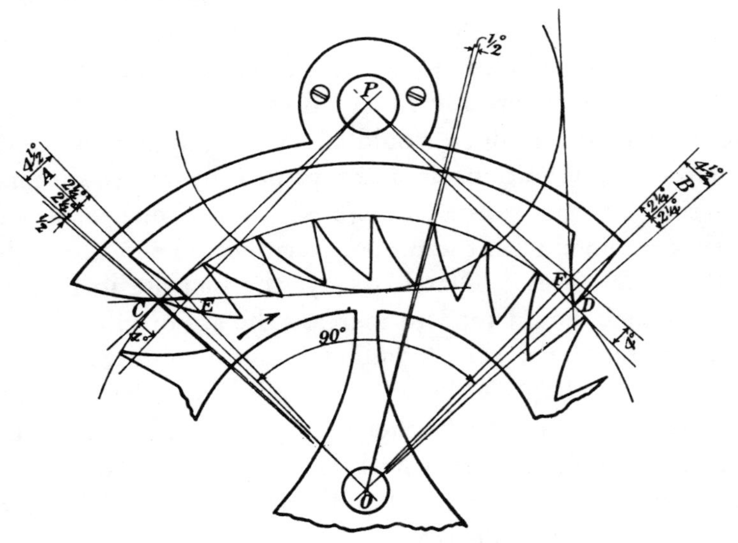

FIG. 20.—The recoil clock escapement.

The figure shows the pallets embracing an arc amounting to the space of 7½ teeth of a ratchet-toothed escape wheel. The wheel has 30 teeth, so that the pallet angle *AOB* is 90°; and the pallet staff centre *P* is given by the intersection of lines *CP*, *DP*, which are tangential to the circle representing the wheel circumference at the points where this circle is cut by *OA*, *OB*. The acting surfaces of the pallets cover an angle of 2¼° on either side of the radii *OA*, *OB*, making each 4½° in all. In addition to this 4½°, 1° is allowed for clearance and ½° for the tips of the teeth, so that the total of 6° makes up the supplementary arc between the 7½ spaces embraced by the pallets and the full amount of 8 teeth. *CPE* and *DPF* are angles of 4° which determine the lengths of the impulse planes. When projected, *CE* and *DF* become tangents to a common circle about *P*, called the "impulse tangent circle." In setting out the escapement, the accuracy of these tangents is a test of the correctness with which the angles have been determined. The length of the recoil, of course, varies with the extent of the arcs of vibration of the pendulum and this fact proves in reality an advantage, because any variation in the motive power which would affect the impulse and tend to alter the vibrations is accompanied by an equivalent variation in the resistance offered at the escape wheel teeth, combined with the proportionate change in the angle of the pallet faces. That is to say, the increase in motive power which would tend to increase the arc of vibration is counteracted by a corresponding increase of force acting against the pallets as the wheel recoils, and, as the recoil increases, there is yet a further resistance occasioned by the increasing angle of the wheel teeth to the pallet faces. The result is that, within certain limits, the arcs of vibration remain very nearly constant.

The Dead-beat Escapement. The dead-beat escapement was introduced by George Graham about the end of the seventeenth century. This escapement is met with in high-class clocks and astronomical regulators, as well as in turret clocks, and its principal feature is that, when a tooth of the wheel falls on the locking, the wheel remains stationary during the completion of the arc of vibration of the pendulum. There is no recoil with its consequent backward motion of the seconds hand.

Fig. 21 shows a plan of this escapement. The pallets

embrace the space of 7½ teeth, as in the case of the recoil, but the locking faces of the pallets are struck from the centre. If a greater number than 7½ teeth are covered, weight is added to the pallets and no advantage whatever is gained for the action of the escapement. The fronts of the teeth of the wheel are undercut about 10° and ½° is allowed for the tip of the tooth. There is 1° from the

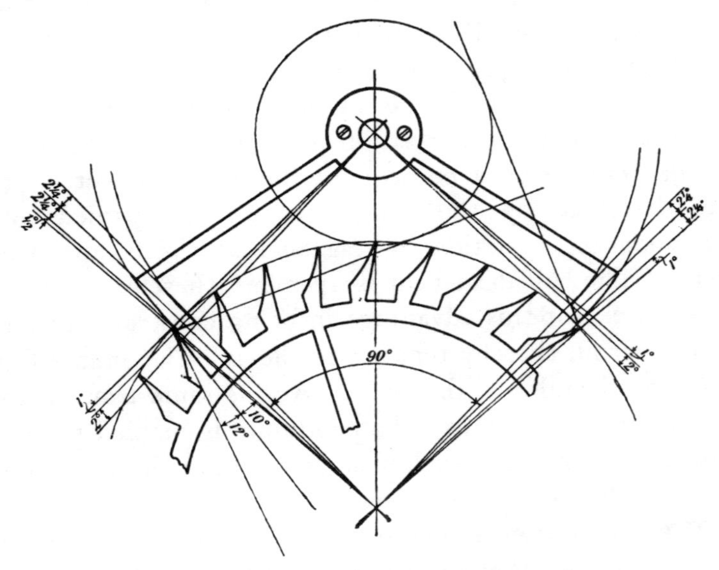

FIG. 21.—The dead-beat clock escapement.

pallet centre devoted to locking and 2° for impulse, and the length of the pallets from the wheel centre is 4½°. The impulse faces of the pallets, as in the recoil, require to be tangents to the impulse tangent circle, as shown, if the escapement is to function correctly. An alternative method of finishing off the backs of the wheel teeth to that shown in the figure is by a straight cut at the same angle of 12° with the front.

Half Dead-beat Escapements. The half dead-beat

escapements referred to in the table are occasionally met with in clocks constructed for use with half-seconds pendulums. In most respects they resemble the perfectly dead-beat, but with this difference, that the locking faces, instead of being arcs described from the pallet centres, are set back in either a curved or straight surface by about 3° from the locking corner. The fronts of the wheel teeth too are generally radial, though they are often set forward slightly in the case of small escapements, thereby increasing the durability and improving the action.

As previously stated, the important feature of the recoil escapement is that any variation of impulse due to deficiencies in the train, with a consequent fluctuating effect on the arcs described by the pendulum, is accompanied by a corresponding variation of resistance at the escape wheel teeth. The half dead-beat escapement, therefore, possesses to some extent this advantage in being constructed with a partial recoil, thereby tending to alleviate the effect of any mechanical faults in the train. At the same time, with a smaller angle of impulse, the pendulum is enabled to maintain its performance within narrower limits and so reduce the trouble arising from circular error.

The Pin-pallet Escapement. The pin-pallet escapement may be either dead-beat or recoil, according to the cut of the wheel teeth. This escapement, invented by Achille Brocot (1817–1878), of Paris, is not suitable for use in high-class clocks, yet it is capable of giving very good results in the French and American marble clocks where it is usually found and frequently operating visibly in front of the dial.

The proportions met with vary considerably, but, having in view the ornamental appearance of the arms, it was usual to make the pallets embrace either 9½ or 10½ teeth. The acting faces of the pallets are semicircular, being cylindrical carnelian stones or pieces of steel shellacked into holes and projecting from the extremities of the pallet arms, with the projecting portion cut half away.

The working of this escapement will be understood from Fig. 22, where two forms of escape wheel teeth are indicated. In one case the fronts are radial when the motion of

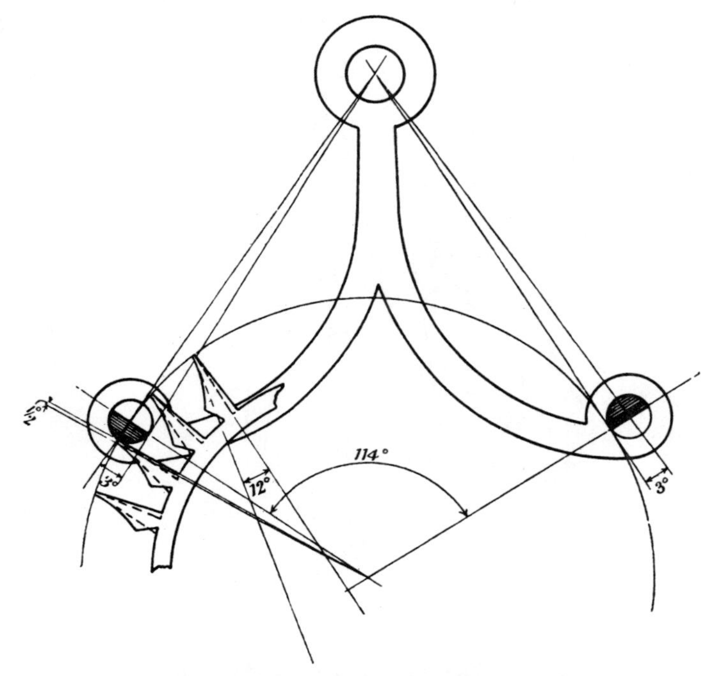

FIG. 22.—The pin pallet clock escapement.

the pendulum will cause a recoil, but the dotted lines show teeth undercut sufficiently to produce a dead-beat action. The diameter of the pallets should be equal to the distance between two consecutive teeth, after allowing a reasonable amount for the width of the tooth and clearance. The impulse resulting from these dimensions approximates 3°.

The Pin-wheel Escapement. Fig. 23 shows the pin-wheel escapement, which is now practically discarded, except for turret clocks. The original invention was by a Frenchman, named Amant, about the middle of the eighteenth century, though his form was somewhat improved upon by J. A. Lepaute (1709–1789). The escape

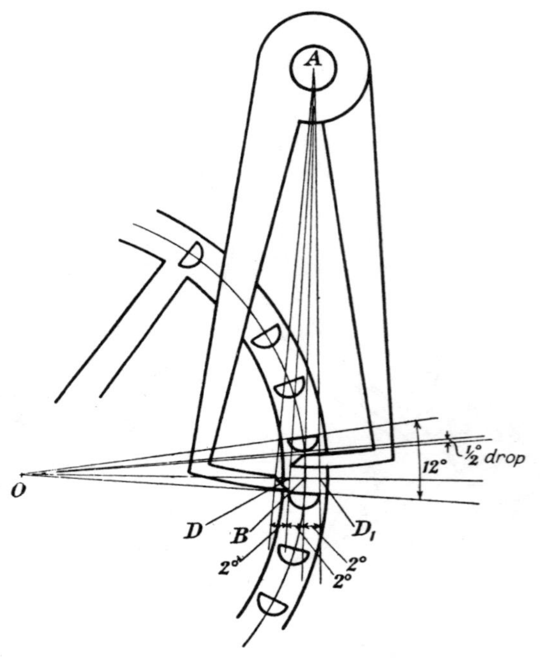

Fig. 23.—The pin wheel clock escapement.

wheel teeth are semicircular pins slightly flattened, projecting from one side of the rim, and the pallets operate tangentially to the wheel as shown. The form of the pallet faces, being arcs taken from the pallet centre, renders the escapement a "dead-beat." In a well-proportioned escapement, the diameter of the pins equals half the distance from one pin to the next. The pallet centre lies on AB, which is a tangent to the wheel radius OB. DD_1 is the diameter or total width of the pins marked off equally on either side of B along OB. 2° are allowed for the escaping arc, so that the complementary angles of 88°, BDA, BD_1A, determine the pallet centre A along AB. The thickness of the pallet arms equals one-half the diameter of the pins and the length of the impulse faces covers 2° from the pallet centre. A tooth is shown just leaving the left-hand pallet whilst the next tooth is about to drop through ½° on to the right-hand pallet.

A turret clock movement fitted with one of these escapements is shown in Plate IV.

The Gravity Escapement. The last and by no means the least interesting of clock escapements to which attention must be directed is the double three-legged gravity type (Fig. 24). It is mainly used in turret clocks, and although several varieties of gravity escapements have been devised, this particular one—introduced by the late Lord Grimthorpe in 1854 for the clock in the Houses of Parliament—bears an exceptionally good reputation. It should, however, be noted that this escapement was in reality a development of an earlier invention by J. M. Bloxam, a barrister, and it was only after some unsuccessful first attempts that Lord Grimthorpe was compelled to adopt Bloxam's idea. Credit, must therefore, be given to Bloxam for a system which has proved so meritorious that it is considered to be the best solution of the remontoire principle ever conceived for larger timepieces. A clock fitted with an escapement by Bloxam can be seen in the Victoria and Albert Museum, South Kensington.

In turret clocks the driving power of the weight is practically constant, but peculiarities in the train and, notably, the effect of rough weather to which the hands may be exposed, cause very serious fluctuations of impulse to the pallets. The feature of this escapement is, however, that the impulse is entirely free from any such disturbing influence. The escape wheel raises an arm which, in

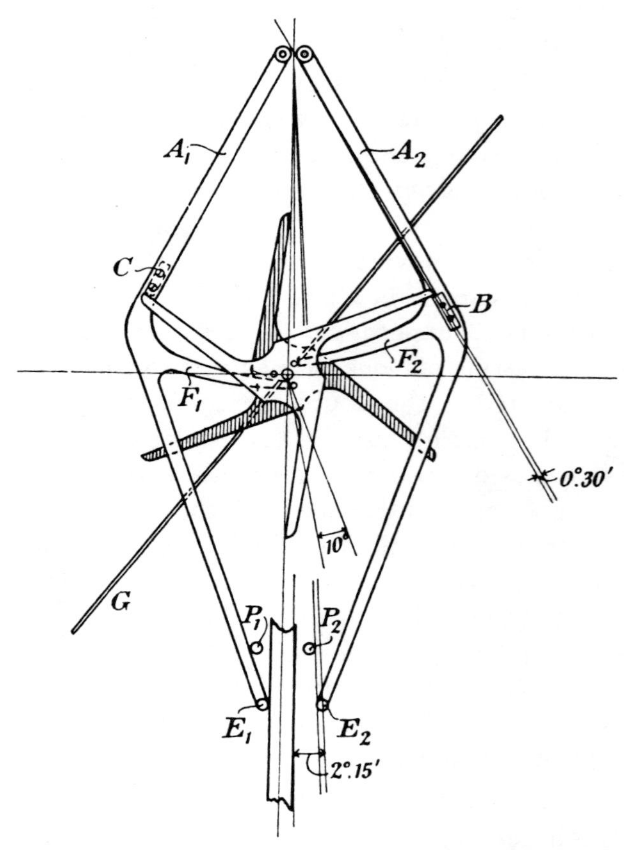

FIG. 24.—The double three-legged gravity escapement as used for the clock in the Houses of Parliament.

falling, rests upon the pendulum rod and provides the necessary impulse by simply increasing the action due to gravity. The pendulum thus receives constant impulse which goes far towards the maintenance of a constant arc of vibration.

In Fig. 24, A_1, A_2 are two arms pivoted as close as possible to the centre or bending line of suspension of the pendulum, that is to say, about one-third of the length below the block, and free to fall by the action due to gravity. The escape wheel consists of two sets of three steel legs separated at the centre by three equally spaced round distance or lifting pins. When mounted, the tips of all these legs are equidistant apart and, in motion, they lock alternately on the blocks *B* and *C*. In the figure, *B* is the front block and *C* is at the back. From the arms A_1, A_2 there are side projections or pallets F_1, F_2, the flat ends of which lie in the path of the round lifting pins.

A leg is shown locked at *B*, the arm A_2 having been raised and held in this position through the action of one of the lifting pins against the pallet, F_2, at the previous advance of the escape wheel. An angle of half a degree is allowed for locking, and the pendulum, represented at the middle of its beat, has a perfectly free run on swinging to the right until it reaches the tail, E_2. This the pendulum raises and, in so doing, the locking at *B* is released. The escape wheel then commences to rotate and the lifting pin nearest the pallet, F_1, passing through an angle of 10°, lifts the gravity arm to the set position for giving impulse and also prepares the block, *C*, for locking accurately the succeeding leg. The pendulum having, at this stage, reached the limit of its swing to the right returns and then receives impulse from the gravity arm, A_2; falling with it. This continues until the pendulum arrives on the line of centres, when the gravity arm is stopped by the banking pin, P_2, and the pendulum again travels freely to repeat the process on the left-hand beat. In the figure, impulse has just been completed from the left-hand gravity arm, A_1; the banking pin, P_1, prevents any further movement of A_1 and the pendulum is directly crossing the line of centres. To equalise the motion of the escape wheel and the force with which the legs drop on the locking, a fly, *G*, is fitted, friction tight, on the escape wheel arbor, operating in a similar way to the fly which terminates a striking train.

The good reputation of the Westminster clock must in a large measure be ascribed to this escapement, coupled with the fact that it is used with a 2 seconds pendulum. Wear on the locking pads is severe and it is essential for the escapement to operate slowly, otherwise the locking may fail and tripping occur. It is, therefore, important that the arc of vibration shall be reduced as far as possible, and better results are usually obtained from 1½ or 2 seconds pendulums than from 1 second pendulums.

It is interesting to record that the finest performance of the Westminster clock shows a mean daily rate of 1 second and, in one year on record, the error was not allowed to exceed 4 seconds without being corrected.

CHAPTER V

TRAINS, MOTION WORKS AND GEARING

The Train. The " train " of a clockwork mechanism has an important part to play. It is the means of communication, by a series of toothed wheels, between the driving force of a weight or mainspring and the escapement. Simultaneously with the transmission of driving force, the speeds of the moving parts are distributed in such a way as to differentiate between minutes, seconds or any convenient intervals of time. The series of wheels alternates between high numbers and low numbers of teeth in order to increase the respective speeds of rotation. Wheels which possess fewer than twenty teeth are described as pinions of so many leaves.

In an ordinary clock, the train consists of a " great wheel," which comes under the direct influence of the driving force, followed by a " centre pinion and wheel," calculated to make one revolution per hour. In some instances these have to be separated by an intermediate pinion and wheel in order to prolong the duration of the driving force. The centre wheel engages with the " third " pinion and wheel, which in turn drives the " escape " pinion and wheel. If the clock has to indicate seconds, the escape pinion is arranged to rotate once in a minute and the arbor is projected through the dial to carry a seconds hand.

The " motion work " is an auxiliary system of gearing to enable a wheel to rotate at one-twelfth or sometimes one-twenty-fourth of the velocity of the centre pinion arbor, by which it is controlled, for the purpose of recording hours.

From previous reference to the behaviour of the pendulum and escapement, the importance of maintaining uniformity of driving force throughout the train will be readily realised. Also, where so many moving parts are involved, friction must be reduced to an absolute minimum, otherwise the requisite driving force has to be increased and wear becomes serious. The greatest care, therefore, must be taken in determining the most suitable arrangement and form of gearing.

When two toothed wheels are in correct engagement, the number of revolutions made by the follower to one turn of the driver is equal to the number of teeth in the driver divided by the number of teeth in the follower. In the case where there is a series of drivers and followers the product of the former is divided by the product of the latter to determine the number of revolutions made by the last follower to one of the first driver. In the train of a clock, the escape pinion must make 60 turns to one of the centre wheel if the escape pinion arbor has to carry a seconds hand, so that in the form of an equation we may put it thus * :—

$$\frac{\text{Centre wheel} \times \text{Third wheel}}{\text{Third pinion} \times \text{Escape pinion}} = 60$$

To go a step farther, we may say that :—

No. of vibrations of the pendulum per hour

$$= \frac{\text{Centre} \times \text{Third} \times \text{Escape wheels} \times 2}{\text{Third} \times \text{Escape pinions}}$$

* This and the following equations are written with abbreviated terms as generally employed in practice. For the sake of completeness, and to make matters clear to students, it should be noted that Centre wheel × Third wheel means the product of the *number of teeth in* the centre wheel multiplied by the *number of teeth in* the third wheel; similarly, Third pinion × Escape pinion means the product of the number of leaves in the third and escape pinions, and so on.

The reason for multiplying the product on the right-hand side by two is because each tooth in the escape wheel covers two vibrations of the pendulum. The foregoing expression is usually abbreviated in the following way,

$$\text{Vibrations per hour} = \frac{2CTE}{te} \quad . \quad . \quad . \quad . \quad . \quad (1)$$

EXAMPLE.—A clock with a pendulum beating seconds has an escape wheel of 30 with a pinion of 8 and a centre wheel of 64. What should be the numbers in the third wheel and pinion?

The pendulum beats seconds, so that the vibrations in an hour are $60 \times 60 = 3600$. Eqn. (1) then becomes :—

$$3600 = \frac{2 \times 64 \times T \times 30}{t \times 8}$$

whence

$$\frac{T}{t} = \frac{3600 \times 8}{2 \times 64 \times 30}$$

$$= \frac{60}{8}.$$

The numbers of teeth in the third wheel and pinion should therefore be 60 and 8, respectively.

Again, the centre arbor performs its revolution in one hour, so that if the clock is to be maintained for so many days and the barrel makes n turns, then,

$$\frac{n \times \text{Great wheel}}{\text{Centre pinion}} = 24 \times \text{Number of days} \quad . \quad . \quad (2)$$

Should intermediate wheels be necessary for reducing the size of the great wheels and prolonging the activity of the driving force they are introduced into the equation as additional drivers and followers.

For motion works, the equation becomes :—

$$\frac{\text{Hour wheel} \times \text{Minute wheel}}{\text{Minute pinion} \times \text{Cannon pinion}} = 12 \quad . \quad . \quad (3)$$

That is to say, the hour wheel rotates only once to twelve revolutions of the cannon pinion, which performs in harmony with the hourly revolution of the centre arbor.

In most clocks the cannon pinion and minute wheel are of the same size and number, so that the actual speed relationship is only between the hour wheel and minute pinion.

From what precedes it can be seen that in practice the number of teeth in the various wheels and pinions of a clock can very easily be determined by reference to one or other of the equations (1), (2) and (3).

Form of Gearing. The next important consideration is the correct method of forming or shaping the teeth of wheels and pinions. An ideal would be realised if there could be frictionless transference of the driving force between engaging wheels and pinions, but this is not possible. In the case of a pinion having fewer than 10 leaves, there is additional friction created between the pair because the form of construction demands that engagement must commence before the line of centres. On the other hand, if a pinion has 10 leaves or more, frictional contact occurs only after the line of centres has been passed. These two forms of friction are known respectively as engaging and disengaging friction, or ingoing and outgoing friction. The relative motion, known as the velocity ratio, between toothed wheels is precisely the same as if they were toothless blanks rotating about their centres, without slipping, by contact friction. Supposing that the leader of two such rotating blanks had a circumference measuring twice as much as the follower, then the latter would revolve twice to once of the former. In an indented wheel, the circumference of the corresponding toothless wheel is an imaginary circle called the "pitch circle" and the teeth are formed partly above (addenda or faces) and partly below (flanks) this circle. The measurement of a tooth and space along the circumference of the pitch circle is described as "circular pitch" and it is essential that wheels in gear shall be of equal pitch. Pitch may also be regarded as the length of an arc of the pitch circle derived from dividing the circumference equally into as many parts as the wheel has teeth and, owing to the fact that equal arcs must be passed over in equal times, the ratio of the length of the arcs to their respective diameters must vary as the circumferences to each other. The velocity ratio is inversely proportional to the radii, diameters or circumferences.

The Epicycloid and Hypocycloid. In clockwork, the essential feature of reducing friction to a minimum, combined with the necessity of ensuring perfectly even propulsion, is realised by adopting a system of engagement which provides for the automatic adjustment of the surfaces in contact. Two very important geometrical curves are involved in forming the teeth in such a way as to produce this uniformity of rotation, or a uniform "lead" as it is described, viz., the "epicycloid" and the "hypocycloid."

The epicycloid is a curve traced by a point on the circumference of a circle which is presumed to be rolling upon the circumference of another circle. A hypocycloid, on the other hand, is a curve traced by a point on the circumference of a circle rolling on the inside of the circumference of another.

The addenda of the teeth of a wheel follow the epicycloidal curve and the flanks of a pinion which it drives are hypocycloidal, in each case originating from the same rolling circle.

In the case of the hypocycloid, it happens that if the diameter of the rolling circle is one-half that of the pitch circle, then the path described by a point on the roller becomes a radial line. Now, if the pinion is provided with a radial flank to its leaf, it follows that the addendum of the wheel tooth should be formed by an epicycloidal curve whose roller has a diameter equal to half the pitch diameter of the pinion into which it gears, whilst the pinion addenda are finished off with semi-circular roundings.

In Fig. 25, the case is taken of a wheel of 64 teeth driving a pinion of 8 leaves, therefore the pitch diameters are in the ratio of 8 to 1. The pinion has radial flanks, consequently

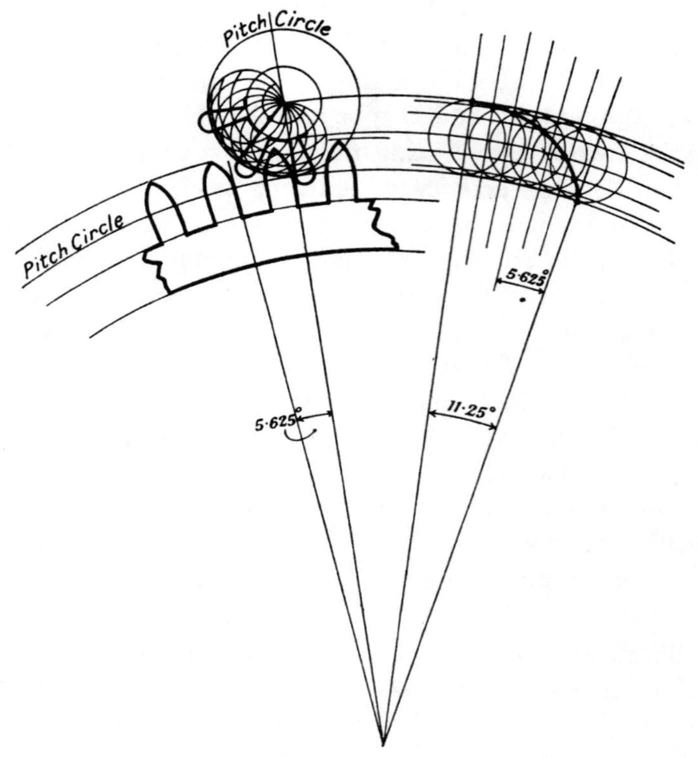

Fig. 25.—Epicycloidal gearing.

the diameter of the rolling circle is one-half that of the pitch circle of the pinion, which produces the hypocycloid shown, as well as the epicycloidal curve for forming the wheel-teeth addenda. The ratio of the wheel to the roller is thus :—

$$64 : \frac{8}{2} = 16 \text{ to } 1,$$

and this represents an angle on the circumference of $\frac{360}{16}$ or 22·5 degrees (22° 30′), for tracing the complete epicycloidal curve, only part of which is required for the addenda. Half the complete curve is shown on the right-hand side of the figure. The pitch of the wheel covers an angle equal to 360 degrees divided by the number of teeth, *i.e.*, $\frac{360}{64}$ or 5·625 degrees (5° 37′ 30″), and is equally occupied by a tooth and a space. The addendum of a tooth above and the radial flank below the pitch circle each occupies half the pitch.

If it be assumed that the wheel has a pitch diameter of 64 mm., then the pinion diameter is 8 mm., and the pitch of the wheel is the pitch circumference (πd) divided by the number of teeth, 64, thus,

$$\frac{3 \cdot 14 \times 64}{64} = 3 \cdot 14 \text{ mm.}$$

If the addendum, as has been shown, is to be half the pitch in height, then the pitch diameter must be increased by the height of two addenda, one on each side, to give the outside diameter of the wheel. It is on the circumference about this diameter that the two epicycloidal curves, approaching from opposite directions, meet.

It is customary for the wheel pitches to be occupied equally by a tooth and a space, though in pinions the space generally takes up two-thirds and the leaf one-third. In the former case, therefore, the outside diameter of the wheel, *i.e.*, the diameter of the blank before being cut, is given by :—

$$\text{pitch diameter} + \text{circular pitch.}$$

In the case of the pinion, if used as a driver with fewer

leaves than 10 and epicycloidal addenda, the outside diameter is :—

$$\text{pitch diameter} + \frac{2\,(\text{circular pitch})}{3}$$

but if the pinion is a follower, with semi-circular roundings, the outside diameter is :—

$$\text{pitch diameter} + \frac{2\,(\text{circular pitch})}{6}$$

In gearing of epicycloidal construction, it follows that for the engagement or " depth " between two wheels to be correct the intersection of the respective teeth on the line of centres must take place also on the pitch circle.

Sometimes, in the case of very large wheels, such as those used in turret clocks, the curves of the addenda, instead of being allowed to meet in a point, are left blunt. The tips are not necessary, as the second tooth has taken up the lead much before the first has reached its limit.

The roots of wheels and pinions are most frequently parallel to the circumference with sharp corners, though often, particularly in drivers, rounded bottoms are preferred as adding strength to the tooth.

Lantern pinions, the leaves of which are formed from round steel pins mounted between two discs, are often used in the going train of turret clocks and also in inferior foreign domestic clocks. The pitch circle passes through the centres of the pins, which have a diameter of rather less than half the circular pitch. To arrive at the suitable epicycloidal curve for forming the wheel addenda in this case, the diameter of the roller should be equal to half the difference of the pinion's pitch diameter and the diameter of a pin.

Involute Gearing. Another curve used in the formation of wheel teeth is the involute. This is the curve traced by the end of a piece of string when unwound from a cylinder. Plotted geometrically, as shown in Fig. 26, and commencing from a given point, *A*, on the circumference of a circle, the curve is determined by a succession of tangents, the length of each being equal to that of its respective arc measured from *A*. Thus, the scale *CD* is equal to the length of the arc *AB*, and both are divided into a corresponding number of equal parts, 1, 2, 3, etc. From the points 1, 2, 3, etc., on the circumference, tangents

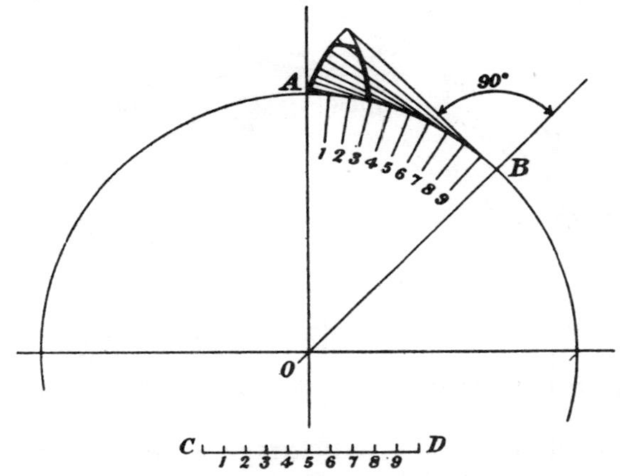

FIG. 26.—Involute gearing.

are drawn towards the radius *OA* equal respectively to the lengths of the scale *C*1, *C*2, *C*3, etc.

The circle upon which the curve is formed is the base circle. In gearing of this kind, engagement is correct whether deep or shallow, provided that the wheels are of equal pitch and acting on a common line of contact. The path of contact may, therefore, vary within certain limits, but the usual inclination from the common tangent at the point of contact of two pitch circles is 15°, known as the pressure angle.

The path of contact is indicated by the line *AB* in Fig. 27, where two pitch circles are shown giving a 2 : 1 ratio.

The line *AB* is then a tangent to the respective base circles. The roots of the teeth are determined by practice and according to requirements.

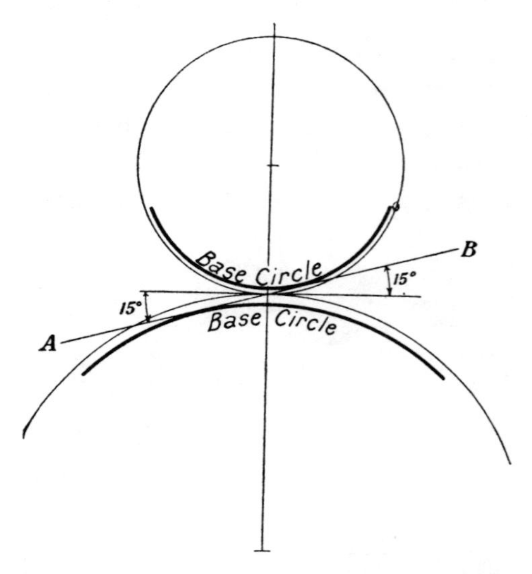

FIG. 27.—Showing inclination of path of contact in Involute gearing.

Strength of teeth is an advantageous feature of involute gearing, and this fact renders it useful for such parts as winding mechanisms, turret clock trains and barrel wheels in certain cases, both in clocks and watches.

CHAPTER VI

WEIGHT-DRIVEN CLOCKS

Weights and Lines. The expression "weight clocks," as the term implies, distinguishes those mechanisms which rely for their driving force on the energy derived from the action of gravity upon a raised mass. It has already been shown that the principal object of a driving force is to maintain the motion of the pendulum, but it has also to provide the necessary power to overcome friction in the train. In order to minimise pressure and consequent wear on the pivots of the barrel or drum upon which the line is wound, it is usual to arrange for the weight to be as light as possible. It has, however, to exert its pull through some definite period, eight days for example, and its length of fall has to be limited. If a weight could be allowed an unlimited fall, it would continue, but for the more or less negligible extra weight of the line, to exert the same pull and consequently keep the train in motion for an unlimited period.

The arrangement of the line may be by merely a single length supporting the weight directly from the barrel edge, or it may be doubled or trebled as shown in Fig. 28. One or other of these arrangements is chosen, generally with the amount of available fall in mind, but it is most important to remember that a weight which would develop a certain force at the barrel edge if supported by a single line must be doubled or trebled if required to develop the same power at the barrel edge when the number of lines is increased correspondingly. Thus, in the figure, the force at the barrel dge of the respective systems is W_1, $\frac{1}{2}W_2$, $\frac{1}{3}W_3$, $\frac{1}{4}W_4$. The arrangement of lines also has a corresponding effect on the length and rate of fall. The weight supported by a single line will keep a clock going throughout a certain length of fall; with a double line, the clock is maintained twice as long for the same amount of fall and the rate of fall is halved; and so on.

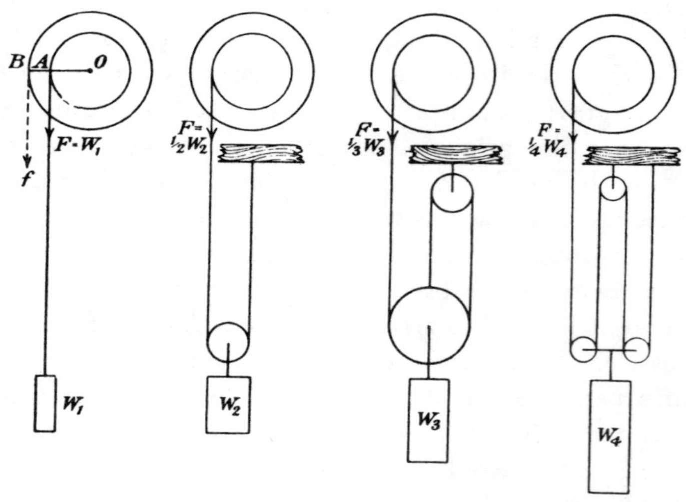

FIG. 28.—Showing various methods of supporting a clock weight.

The mechanical "principle of work" enters widely into calculations concerning appropriate weights for clocks. This principle is summed up in the axiom: "What is gained in force is lost in speed and what is gained in speed is lost in force." The application of this principle to a clock manifests itself in maintaining the motion of the pendulum by means of a small force delivered from a relatively fast moving part. This small force is the outcome of a series of varying forces and speeds from the original motive power which is slow motion at high pressure. At any stage of this progress the actual amount of work done is the same, but for the loss through friction, because as the force is reduced the distance traversed by the acting part is proportionately greater for any given time.

To ensure that this condition prevails throughout the train, it is essential that the moments of the forces at the acting points should be equal. The moment of a force is the tendency of the force to rotate a body to which it is applied and is the product of the magnitude of the force and its leverage or its perpendicular distance from the axis of rotation.

In Fig. 28 the moment of the force acting on the barrel edge at A equals the distance or radius, OA multiplied by the weight, w, but at the great wheel teeth, B, the moment of the force will be $OB \times f$, so that if OA is two-thirds of the distance OB, then the force, f, acting at B will amount only to two-thirds of the weight, w, in order that the moments may remain equal.

Thus,

$$\text{Moment at } B = \text{moment at } A$$

that is, $$f \times OB = w \times OA$$

but, $$OA = \tfrac{2}{3}OB$$

hence $$f = \tfrac{2}{3}w.$$

The actual amount of work done (= Force × Distance) is the same in both cases, because the distance traversed by the great wheel teeth is proportionately greater than that traversed by the barrel edge during a given time.

Thus, the work done, W, for one revolution of the barrel is equal to the weight, w, multiplied by the circumference, πd, but for one revolution of the great wheel the circumference is $1\frac{1}{2}$ times greater and the force is $\frac{2}{3}w$, so that the work done is the same.

Throughout a clock train, the force is applied at the

smallest radius so that at all the acting points the force is reduced but the distance increased, yet, excepting for loss by friction, the work done is everywhere equal to that performed by the weight.

The formula for computing the length of fall of a weight or the number of turns made by a given barrel, as the case may be, is as follows :—

$$N = \frac{L}{\pi(d + t)},$$

where N is the total number of turns of the barrel; d, the diameter of the barrel; L, the length of line; and t, the thickness.

EXAMPLE. What would be the necessary diameter of a barrel making 16 turns, if the case admitted a fall of 6 ft. 6½ ins. and a double line $\frac{1}{16}$ in. thick were to be used?

From the above formula:

$$d = \frac{L}{\pi N} - t$$

$$= \frac{78 \cdot 5}{3 \cdot 1416 \times 16} - \frac{1}{16}$$

$$= \frac{78 \cdot 5 - 3 \cdot 1416}{50 \cdot 2656}$$

$$= 1 \cdot 5 \text{ ins. (approx.).}$$

Maintaining Works. Before passing from this subject, reference must be made to a device known as the "maintaining power." When a weight has come to the end of its fall and requires to be rewound, the action of winding produces a tendency to reverse the motion of the train and drive the clock backwards. The maintaining power averts this tendency and keeps the mechanism going whilst being rewound. In practice, the system adopted for this purpose may differ slightly in form, but the usual application is by means of a click or "detent" pivoted between the frames acting on a ratchet wheel mounted between the great wheel and barrel ratchet on the winding arbor. The great wheel and maintaining ratchet are free on the arbor but, on the drum side of the maintaining ratchet, a click is provided to engage the barrel ratchet. The maintaining spring is circular and fits into a recess in the great wheel; one end is made fast to the great wheel and the other end is attached to the adjacent maintaining ratchet. When the winding commences, the backward movement of the maintaining ratchet is immediately checked by the action of the pivoted detent, which winds the maintaining spring, whereby the great wheel is able to continue its forward motion.

In turret clocks, various forms of maintaining devices are met with. Sometimes a weighted lever when at rest is arranged to cover up the winding arbor or otherwise prevent a key being inserted until it is raised out of the way. This operation causes a click or pawl, mounted on the same arbor as the lever, to drop into one of the leaves of a train pinion. The space between the two leaves must be sufficient to allow clearance, and the weight of the lever must be great enough to keep the power on during the process of winding. An illustration of this is shown in Fig. 31.

Weight Clocks. Weight-driven clocks may be regarded as occupying three principal groups: (1) Regulators, bearing the highest attributes of workmanship and skill, which are used in astronomical observatories and places where the greatest accuracy in timekeeping is essential. (2) Long-case, or grandfather clocks, the familiar domestic mechanisms, specimens of which are to be seen dating far back into the early ages of the craft. (3) Turret clocks, which are built into the towers of churches, public buildings, etc., for the benefit of the community in general.

Typical examples of each of these classes are shown in the succeeding diagrams, which will make their construction clear. It should be remembered, however, that although the general principles are maintained throughout any horological mechanism, whether it be a clock or a watch, yet the constructive application may be slightly varied in ways too numerous to record, and therein lies the skill of the expert to recognise and accommodate himself to peculiarities rather by experience than by direct instruction. It is of the utmost importance, therefore, for the student to master fully the fundamental principles of the subject, after which he can be sure of making an intelligent survey of any mechanism with which he may be confronted.

Regulators. Fig. 29 shows the front and side elevation of a regulator movement. In modern forms of construction, the whole mechanism is carried on a cast-iron bracket, *A*, secured to the back of the case. The suspension block from which the pendulum hangs is incorporated in this casting. The usual motion work is dispensed with as a means of reducing friction in the mechanism, and the hour hand operates on a separate dial circle similar to the seconds circle. The three long pivots to carry the hands are shown in the figure; seconds, *S*, on the escape wheel arbor; centre, *C*; and hour, *H*, respectively, the seconds pivot and the hour pivot being equidistant from the centre. The hour wheel which is mounted on the arbor, *H*, may be driven directly from the centre pinion or from the great wheel but it may also be driven off a wheel screwed concentrically to the back of the great wheel, as illustrated. The line is guided over a loose pulley, *P*, so that the weight may fall close to the case and as far away from the pendulum as possible. This is done because in falling and approaching the level of the bob the weight itself becomes sufficiently pendulous to be influenced by the motion of the pendulum if too near and the performance of the clock would be affected.

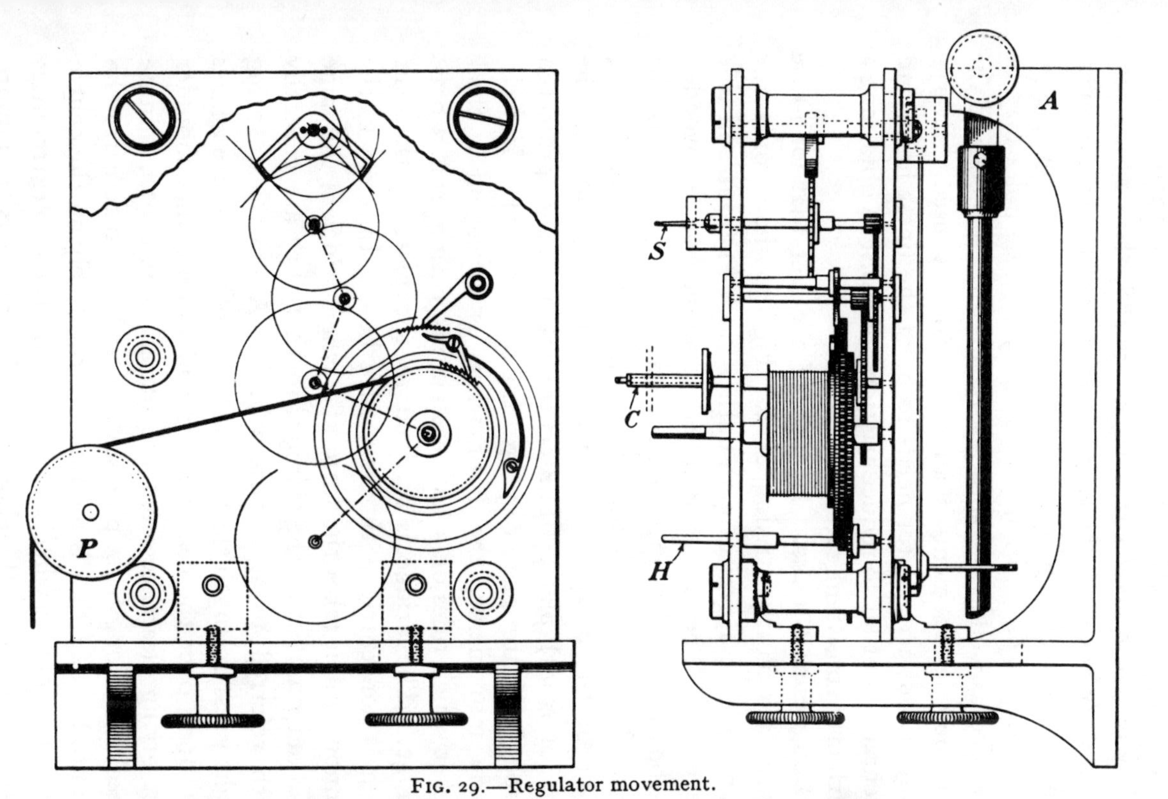

FIG. 29.—Regulator movement.
Left.—Front elevation with the top plate removed showing the arrangement of the train.
Right.—Side view showing the cast iron bracket from which the pendulum is suspended.

A high numbered train is customary in the finest work, as follows: *great wheel* 192 *teeth, centre* 128, *third* 120, *escape* 30 *and pinions of* 16 *throughout. Pinions of* 12 are, however, sometimes used in which case the numbers become: *great wheel* 144 *teeth, centre* 96 *and third* 90. The holes may be jewelled as well as the acting surfaces of the pallets and the dead-beat escapement is almost invariably used. Everything about a regulator needs to be of the finest finish and accuracy in order to reduce friction and assist in obtaining a perfectly consistent rate.

Long-case Clocks. In view of the fact that long-case clocks are mostly striking clocks, Fig. 30 shows only the going-side of a typical grandfather clock in front and side elevation. The movement is mounted on a wooden seatboard which is supported and screwed to the sides of the case. If a seconds hand is carried, it is fitted on a long pivot of the escape wheel arbor projecting through the top (or front) plate. The centre arbor is also projected through the top plate to propel the motion work. In well-made movements, the hour wheel is keyed on a thick socket which rides on a hollow stem projecting from a brass bridge. The bridge is screwed to the plate so that the stem is concentric with the centre hole. In a clock, the cannon pinion takes the form of a wheel which is the same size and number as the minute, so that they are generally both referred to as minute wheels. A brass leaf spring slightly less in length than the diameter of the minute wheels is slipped over the centre arbor and caught on a shoulder just giving clearance from the plate. The centre minute wheel has a long brass pipe squared at the end to carry the minute hand. This is mounted to run freely on the centre arbor, the minute wheel itself resting on the two extremities of the leaf spring. The hour wheel bridge clears the minute wheels and the stem passes over the pipe of the centre one.

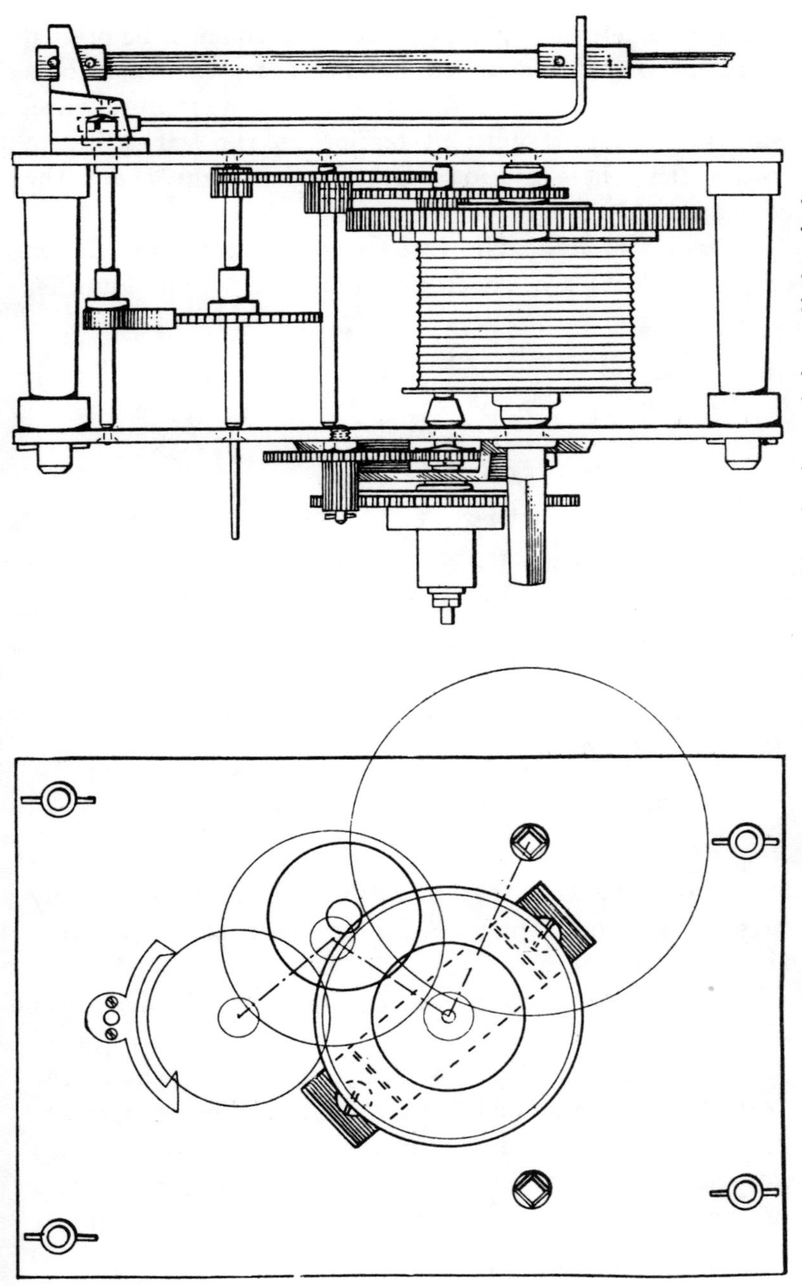

FIG. 30.—Long-case movement. Showing the going side only of a typical grandfather clock.

This pipe is left long enough to enable a collet to be pressed over the boss of the minute hand so that, when held by a pin through the end of the centre arbor, the minute wheel will have put just sufficient tension on the leaf spring to render the whole motion work friction tight about the centre arbor.

Turret Clocks. The varieties in the form and design of turret clocks are extremely numerous. This doubtless

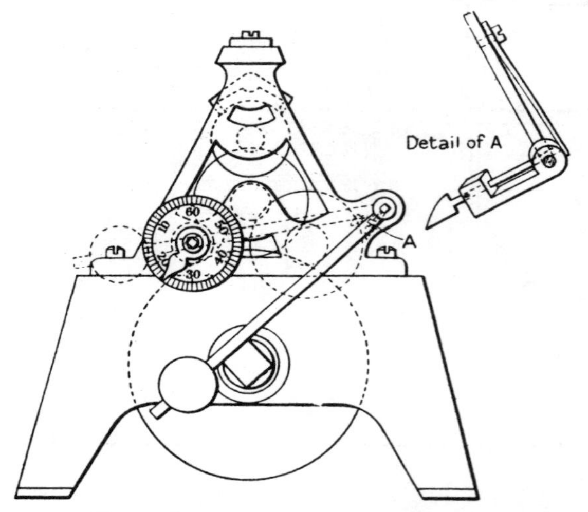

FIG. 31.—Turret timepiece movement.

arises from the fact that many have had to be suitably arranged and built into towers or other places differing widely in architectural form.

Modern turret clocks are usually fitted with the dead-beat or the pin-wheel escapement; the double three-legged gravity escapement is, however, employed where the greatest degree of accuracy is required. Difficulties affecting the behaviour of turret clocks arise from one or two exceptional causes. The exposure of the hands to wind and snow often leads to trouble and the question of counterpoising has to receive very studied consideration. Draughty towers often play havoc with the freedom of the pendulum. Rusting of steel lines, especially if these lead the weights away over pulleys for any considerable distance, increases friction to such an extent that the power falls off.

Among the different designs of towers, and as a kind of general classification, the following may be mentioned. The larger varieties have four dials together with an hour bell and four quarter-chiming bells. The smallest variety is a simple timepiece with one dial which may be arranged either as a dormer in a roof or against a wall; whilst another type, of intermediate size, has two or four dials, striking hours only, and is suitable for public buildings or stables. It is of practical advantage to ensure where possible that the bevel work, which forms the point of transmission to the separate motion works, is planted centrally with the dials, and that the connecting rod between the going train and the bevel work is vertical.

Among small turret timepiece movements Fig. 31 may be taken as a typical example. The frame is cast-iron, bushings and wheels are gun-metal, and a maintaining work, shown in detail, is provided.

A description of other turret movements appears later in Chapter VIII, on Striking and Chiming Mechanisms.

CHAPTER VII

SPRING-DRIVEN CLOCKS

Springs and Barrels. In the category "spring-driven" clocks are placed the numerous types which derive their motive power from the energy of a coiled mainspring. Experience has proved that hardened and tempered steel is the most desirable material for this purpose on account of its superior elastic property. Above all, it is most important that mainsprings should offer resistance to bending without remaining "permanently set" when the stress is removed. That is to say, in coiling a mainspring into a barrel it must undergo distortion well within its limit of elasticity, so that, in unbending, it will return to its original form. During the process of manufacture mainsprings remain perfectly straight strips until the last operation in the factory, which is to coil them into wire rings of suitable diameters. When a spring is removed from a wire it is found to have assumed the form of a coil, and this condition of set should be final, so that, whether the spring is subsequently coiled or uncoiled, it should not vary from this form when relaxed. Unfortunately, in practice, springs vary very considerably in their behaviour. Some soon become exhausted, losing their vitality, while others, for some unaccountable reason, possibly a rapid change of temperature affecting a weak spot, suddenly break. Inferior metal, careless tempering and bad fitting are very frequent causes of trouble.

Space will scarcely permit a close investigation of the dynamical formulæ governing the development of the coiled spring as a motive force. This part of the subject is more closely dealt with in its bearing on balance-springs given in Chapter XIII. To the clockmaker, the principal question is : What should be the correct dimensions of a spring to suit a particular barrel? The centre portion of the barrel is occupied by the arbor and the remaining orifice has to contain the spring whether wound or unwound. Therefore, in order that the spring may develop the maximum number of turns of winding in the barrel it must occupy half that available space. An arbor generally has a diameter one-third that of the barrel in cases where few turns of winding are needed, but this can safely be reduced to one-quarter for clock barrels making eight turns.

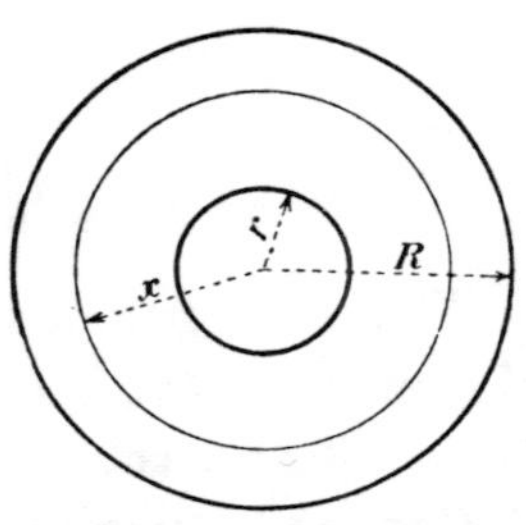

FIG. 32.—Showing the relative positions occupied by a coiled spring and arbor in barrel.

A glance at the illustration, Fig. 32, will make the following formula clear :—

Let R = radius of the barrel,
and r = that of the arbor.

Then, the area available for the spring to occupy is—

$$\pi R^2 - \pi r^2, \text{ or } \pi(R^2 - r^2)$$

The circumference of a circle which divides this area into two equal parts lies nearer to the edge of the barrel than to the arbor.

Let the radius of this circle be x, then—

$$2(\pi x^2 - \pi r^2) = \pi(R^2 - r^2)$$

whence $$2(x^2 - r^2) = R^2 - r^2$$

and $$x^2 = \frac{R^2 + r^2}{2}$$

therefore $$x = \sqrt{\frac{R^2 + r^2}{2}}$$

The number of turns of winding, N, is equal to the difference between the number of coils occupied by the spring unwound and wound. If t be the thickness of the spring, then Nt must be equal to the difference between the width of the space from the arbor circumference to the circle whose radius is x and that from x to the barrel edge. Hence—

$$Nt = (x - r) - (R - x)$$

and

$$t = \frac{2x - (R + r)}{N}$$

The two principal methods of applying mainsprings in clockwork for distributing the power are known as the "fusee" and the "going barrel."

Fusees. In this case, the barrel containing the spring is simply a cylindrical receptacle, the outside surface of which is perfectly smooth. One end of the spring is attached by a hook on the arbor and the other by a hook in the barrel edge. It will be easily understood that when a spring is fully wound it will tend to exert greater force than when nearly unwound, so that there is a variation in the pull throughout the whole operation of unwinding. With a view to equalise this natural variation, with its consequent erratic effect upon the escapement and time-keeping generally, the fusee was originated. This ingenious device consists of an arbor whose edge or periphery is turned into the form of a pre-determined hyperbolic curve and then cut with a spiral groove to conduct a chain or gut line connecting the barrel. This curve corresponds to the moments of the varying forces exerted by the spring, which are the product of the force and the leverage, or acting radius, at all the stages of unwinding. As the pull of the spring weakens the larger radii of the fusee are presented to the barrel, the leverage increases and the moments are thus equalised. Conversely, when the spring is fully wound and at its strongest the action is at the smallest fusee radius and the leverage shortest. By this means an even distribution of power is maintained as far as possible. The barrel arbor does not rotate; but in winding the fusee, the chain is pulled off the barrel and the mainspring within is coiled around the arbor in the process. When the chain is unwound from the fusee, it is kept taut with the barrel because the mainspring is "set-up" or coiled slightly into tension. The arbor is projected through the frame and squared to carry a ratchet which is held in position by a click screwed to the frame. The usual method adopted to prevent overwinding consists of a nose provided at the small end of the fusee which butts against the extremity of a lever pivoted to a slotted cock on the frame. This lever is delicately sprung to allow a limited up-and-down action and is planted tangentially to the fusee-steel. Towards the completion of winding the chain comes into contact with the surface of the lever, raising it into position for the nose to butt at the finish of the turn. Normally, the nose passes beneath the lever.

Further details relating to the theory of the formation of the fusee are given in Chapter XV on the subject as applied to watchwork.

Maintaining powers similar to those described in Chapter VI for use with weight-operated barrels are introduced in all well-made fusees. The thick part of the fusee, known as the fusee brass, is recessed and the ratchet is screwed or pinned therein. The maintaining wheel carries two clicks with circular springs to each and, when mounted, these run clear in the fusee brass recess to engage with the ratchet. Thus the fusee brass and maintaining wheel, as well as the great wheel, are brought up flush with each other on the arbor and then keyed on as in the former case.

The origin of the fusee has for many years been attributed to Jakob Zech of Prague about the year 1525. This theory, however, has been disposed of as a result of the investigations which have been made concerning the history of the magnificent table clock of Gothic architectural design made for Philip the Good of Burgundy (1419–1467). This clock was formerly in the possession of Prince Eduard von Collalto of Vienna, but was acquired in 1846 by the Viennese art collector Friedrich Otto Edler von Leber. On his death in the same year it passed to his son, Maximilian von Leber who publicly exhibited it in the Paris Exhibitions of 1878 and 1900 and published in French a short account of the clock, entitled *Notice sur l'horloge gothique construite vers* 1430 *pour Philippe III, dit Le Bon, duc de Bourgogne*, Wien, 1877. Recently it was bought by its present owner, the collector, Mr. Carl Marfels, of Neckargemünd, near Heidelberg and forms the subject of a finely illustrated monograph by Dr. Ernst von Bassermann-Jordan, *Die Standuhr Philipps des Guten von Burgund* (Wilhelm Diebener, G.m.b.H. Leipzig). The fact of particular interest about this well-preserved specimen is that it is provided with two fusees, gut lines and mainsprings, and indisputable evidence shows that both movement and case are original and of the same period and that the clock must have been constructed before the middle of the fifteenth century, probably about the year 1430 or a little later. Thus it is established that the fusee was known fully a century before the time it was hitherto thought to have existed.

Prior to the general introduction of the fusee into watches, a contrivance known as the "Stackfreed" was used to counteract the varying pull of the mainspring. A strong tension spring terminating in a roller acted as a break by pressing on the edge of a snail which rotated as the mainspring unwound. When the mainspring was fully wound the snail had its longest radius in contact with the tension spring, which, in consequence, then exerted its greatest force as a break to the motion of the snail.

Going Barrels. The " going barrel " has no complication at all. The cylindrical box containing the mainspring is mounted actually on one side of the great wheel. The arbor has a square projecting through the frame for the purpose of winding. It will be noticed, therefore, that in winding, the arbor is turned and the barrel remains stationary—just the reverse of the fusee method. Ordinarily, there is no provision for equalising the varying pull of the mainspring in unwinding, but in high-class clocks a " stop-work " is sometimes introduced. By this means the mainspring is set-up and only the middle turns, which exert a more uniform pull, are brought into operation. It is the rule to employ longer and weaker springs in going barrels than in fusee barrels, in order to increase the uniformity of the middle turns as far as possible. Although many clocks without stop-works keep good average time, in most instances, and particularly where the period is 8 days, there is a very pronounced tendency towards variation of rate between the early and latter part of the period. The going barrel stop-work has a " finger " which is fitted usually on a pentagonal or hexagonal projection of the barrel arbor and in rotating engages a slit in a star wheel at every turn. The star wheel is screwed, by means of a shoulder screw, to the cover, or sometimes the great wheel, according to the design. After the finger has made a requisite number of revolutions it is brought to a stop by contact with the surface of one of the teeth in the star wheel which is a full tooth instead of being concentric with the finger as the others. The same thing applies when the motion is reversed but, throughout the whole procedure, the action is locked.

Spring Clocks. Spring-driven clocks are numerous in variety, and may be grouped into three classes, viz., Bracket, Dial and Portable.

Bracket clocks cover a very wide range of specimens for domestic use: antique and modern, English and foreign. Some early types had verge escapements, but for the majority the recoil escapement is invariably used. Apart from the differences in size and, of course, the fact of it being driven by a mainspring, the construction of a bracket clock remains similar to that of a weight clock. In this category, however, come several types distinguished by the manner in which they are cased. There are the fine specimens of English clocks in elegant cases of ebony, walnut or mahogany with glass panels at the side and brass fittings. There are the exquisite French Boulle and decorative mantelpiece clocks. There are "lantern" clocks in cases entirely of brass surmounted by a large hemispherical bell, producing a resemblance to an old-fashioned lantern. There are the more common French marble clocks of designs innumerable, and a number of modern varieties which are in domestic use.

Dial clocks are so termed on account of their plain prominent faces, painted white with black Roman figures and fitted with a pair of spade hands. These hang on the walls of offices, halls and like places. Good 8-day English dial clocks are provided with a fusee movement and Fig. 33 illustrates a typical example. The fusee and gut line can be seen distinctly. Foreign and many modern English "dials" are cheapened by dispensing with the fusee and substituting a going barrel; in fact, a large number of small foreign timepiece movements are now actually fitted into English cases.

Portable clocks, too, are varied in external appearance though fundamentally similar in construction except that they are all provided with a form of watch escapement. They comprise, however, some very antique specimens intended to stand on a table and now rarely to be seen

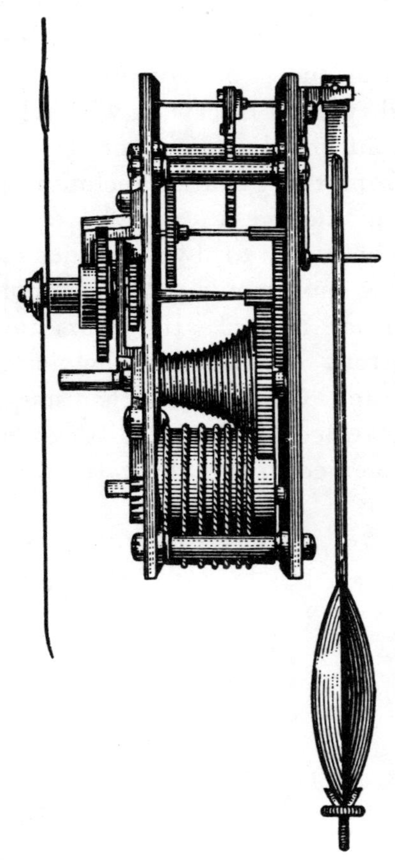

FIG. 33.—Side elevation of an English "Dial" timepiece.

outside museums. These appeared in Europe about the sixteenth century. To see the time one had to look down upon them after the fashion of reading a sundial. Later generations saw the introduction of the sedan clock—a curious timepiece in a circular ebony case, resembling a round picture frame—to hang within a sedan chair. Still later came the ever-popular carriage clock with its lacquered or gilt brass frame and glass sides, front and door, through which the internal construction can be viewed. These movements are now fitted also to small imitation lantern clocks to stand on pieces of furniture.

These are all types of clockwork with which the practical man may be confronted at any time, but there are no real divergencies from the essential technical principles upon which their construction is founded.

It is hardly necessary to dwell further on the design of simple timepiece movements. More important specimens are generally made to strike the hours and many also to chime the quarters. In the following chapter, which deals with Striking and Chiming Mechanisms, the plan and arrangement of wheels and pinions devoted to the going train will be observed in the illustrations.

CHAPTER VIII

STRIKING AND CHIMING MECHANISMS

Striking. The "striking" mechanism of a clock is entirely separated from the "going," the only connection between the two being that the "going" releases the "striking." The latter consists of an independent train of wheels and pinions driven by a separate weight (except in the case of some very antique clocks) or mainspring.

One of two systems, known respectively as "locking plate" and "rack," is invariably met with. The locking plate is applied, in slightly varying forms, to all the earliest types of striking clocks, in turret clocks, early French and even in many modern foreign mass-production clocks. The rack was introduced about the beginning of the eighteenth century and appears in most domestic clocks.

In both systems the train of wheels terminates in a fly, which acts as a governor and produces uniformity in the blows of the hammer.

Locking Plate. Fig. 34 shows the earliest arrangement as applied to long-case clocks for the purpose of striking the hours successively. The minute wheel, *M*, is seen on the point of raising the lifting piece, *L*, which is mounted on the squared projection of an arbor pivoted in the frame. Within the frame this arbor carries the warning lever, *W*, upon which rests the end of a short detent-like lever, *R*, mounted on a second arbor pivoted slightly above the warning lever arbor. Mounted on the second arbor there is, in addition to the lever *R* another

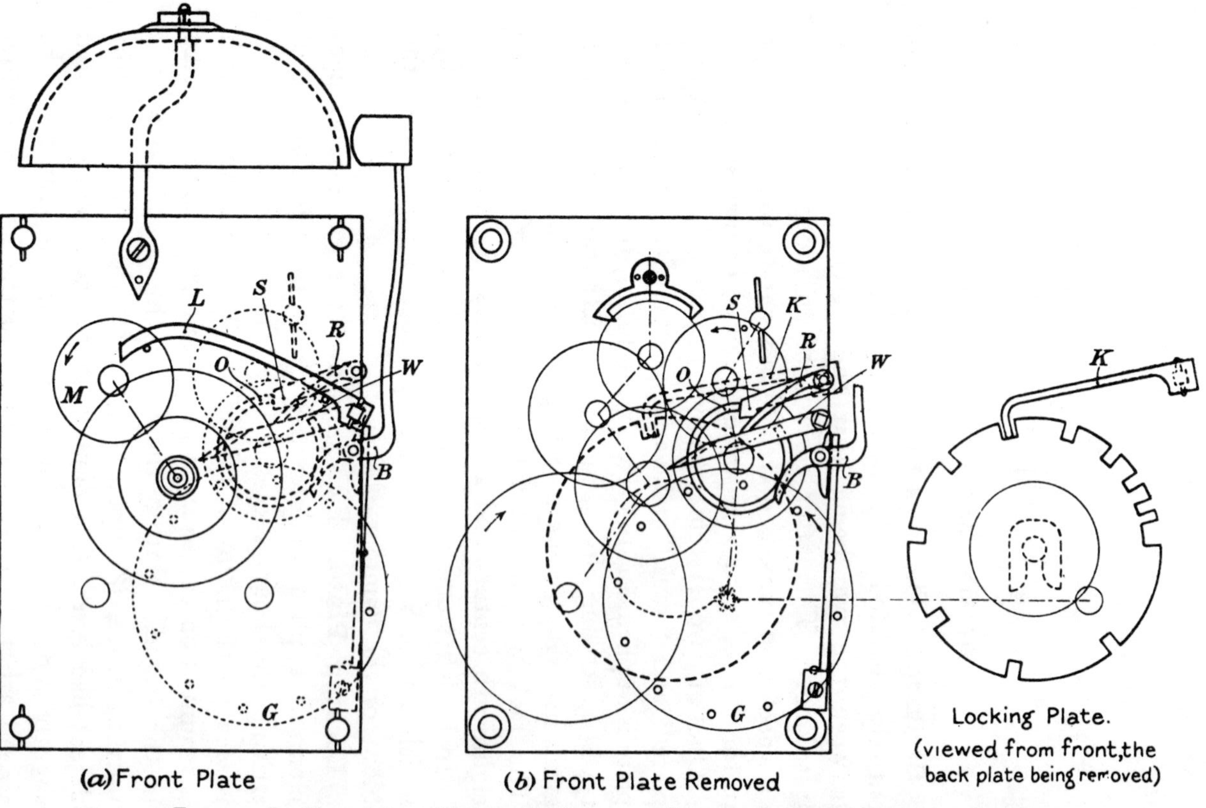

FIG. 34.—Locking-plate striking mechanism, as applied to early English Long-case clocks.

lever, *S*, also within the frame, which acts as a stop to the motion of the train. Again on this same arbor, but on a squared projection beyond the back plate, is mounted the locking lever, *K*. A brass edge or hoop is attached to the side of the wheel next to the great wheel, extending for about $\frac{7}{8}$ths of the circumference, just below the roots of the teeth. This wheel is known as the " hoop wheel " and the train is normally locked by the end of the hoop, *O*, butting against the end of the stop lever, *S*. As the lifting piece, *L*, rises, the warning lever, *W*, the second lever, *R*, and the stop lever, *S*, rise with it as well as the locking lever, *K*, which comes out of its notch in the locking plate. As soon as the stop lever, *S*, clears the butt end of the hoop the train is released. A moment later, it is temporarily brought to rest by a pin on the warning wheel being caught on the extremity of the warning lever, *W*. This extremity is bent so as to move into the path of the pin on the warning wheel as the lifting and warning pieces rise. Precisely at the hour, the pin on the minute wheel passes the tip of the lifting piece, which falls. At the same instant the warning lever, *W*, clears the pin and striking commences. The locking plate determines the number of blows struck, whilst the twelve pins on the great wheel, *G*, operate the tail of the hammer lever, *B*. Until the locking lever, *K*, is allowed to drop into a notch in the locking plate, so long is the stop lever, *S*, compelled to keep clear of the broken section of the brass edge on the hoop wheel. The great disadvantage in the locking-plate system for ordinary clocks is that the position of the locking plate itself is quite irrespective of the going part. The number of blows must perform consecutively for every release of the striking, resulting in chaos if by any chance the clock should be allowed to run down or the hands be advanced in setting the time without allowing the striking to operate appropriately. In turret clocks, however, this objection does not arise to the same extent and this principle is generally adopted, though the application, as also in French clocks and others, is slightly different from the old long-case method described above.

A feature should be noticed in the example illustrated in Fig. 34, namely, the curious form of gearing between the great wheel and the locking-plate wheel. The latter is half the diameter and flush with the locking plate and is driven by a six-toothed end-pinion formed from the projection through the plate of the great wheel arbor. The figure also furnishes an instance of a clock with one weight for driving both going and striking. The rope or chain pulleys are mounted on the great wheel arbors, one of which turns towards the right and the other towards the left as indicated by the arrow heads in the figure. The weight is suspended in a loop formed in the rope or chain between the two pulleys.

Fig. 35 shows the form of locking-plate striking as usually applied to French clocks. The minute wheel, M_1, carries two pins for raising the lifting piece, *L*, at the hour and half-hour. The arm, *W*, which is in one piece with *L*, has a projection near the far end passing through the front plate to operate the warning. As the lifting piece rises, *W* rises too, carrying with it the brass piece, *R*. This is squared on an arbor pivoted in the frame, which carries also the stop lever, *S*. As *W*, *R* and *S* rise together the train is released, temporarily checked by *W*, but finally set in motion by the lifting piece dropping clear of the pin on the minute wheel. The locking plate is mounted on the arbor of the pinion next to the great wheel. This arbor projects through the back plate, as also a knife edge piece, *K*, attached to the arm of the stop lever for the purpose of engaging with the notches in the locking plate.

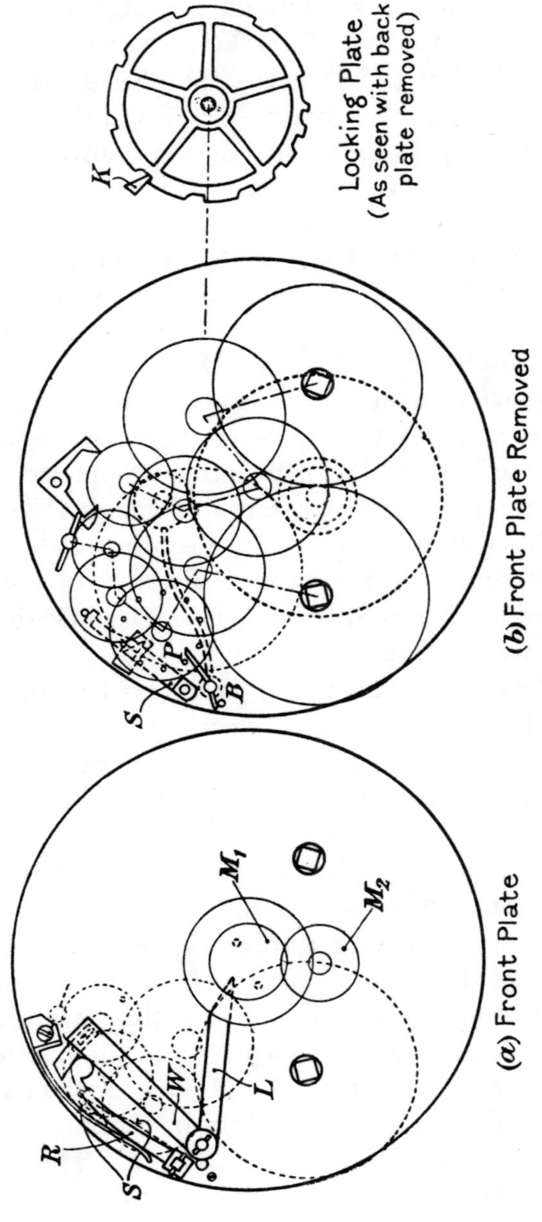

FIG. 35.—Locking-plate striking mechanism as applied to early French clocks.

As long as the knife edge rides on the edge of the locking plate, so long is the stop lever held clear of the path of the pin on the last train wheel. The half-hours are produced by making the locking-plate notches of double width. The stop lever being released allows the pin to pass, but the knife edge drops into the notch again immediately the train proceeds and only one blow of the hammer follows. The pin wheel usually has 10 pins, which raise, in passing, the lever on the pivoted hammer arbor, *B*.

Examples of locking-plate striking as applied to turret clocks are shown in Plates III and IV. The lifting piece is raised by a snail cam on the centre arbor situated behind the minute dial and the locking lever which drops into the notches in the locking plate is mounted on the lifting piece. A detail of interest, supplementary to Chapter IV, on Clock Escapements will be observed in the illustrations. Plate III shows a turret clock movement fitted with the double three-legged gravity escapement whilst Plate IV shows one with a pin-wheel escapement.

Rack Striking. This system has proved to be exceedingly reliable and, for domestic clocks, it would be hard to imagine any way in which it could be surpassed. Virtually the hour wheel, carrying the hour hand, regulates the number of blows struck and, so long as the hour hand is in any one position, only the relevant number can be repeated if the striking is released.

Fig. 36 shows the rack system applied to a long-case clock. Mounted on the hour-wheel socket is a snail, *O*, and impinging on the edge of this is a pin projecting from the tail, *T*, of the rack, *R*. According to the depth of the snail, so is the movement of the rack governed. The lifting piece, *L*, is raised by the pin in the minute wheel, M_1, and in so doing lifts the rack hook, *H*. The rack falls and the number of teeth corresponding with the hour to

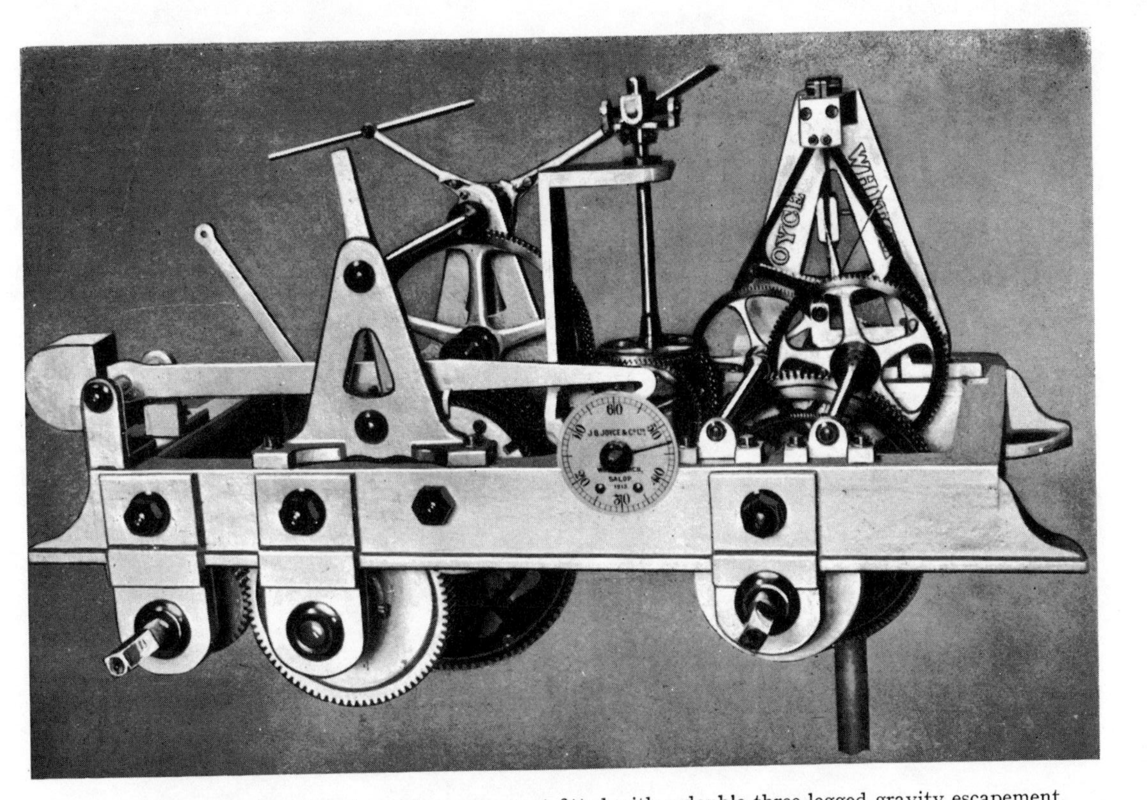

PLATE III.—A modern turret striking movement fitted with a double three-legged gravity escapement.

[To face page 92.

PLATE IV.—A modern turret striking movement fitted with a pin-wheel escapement.

[*To face page 93.*

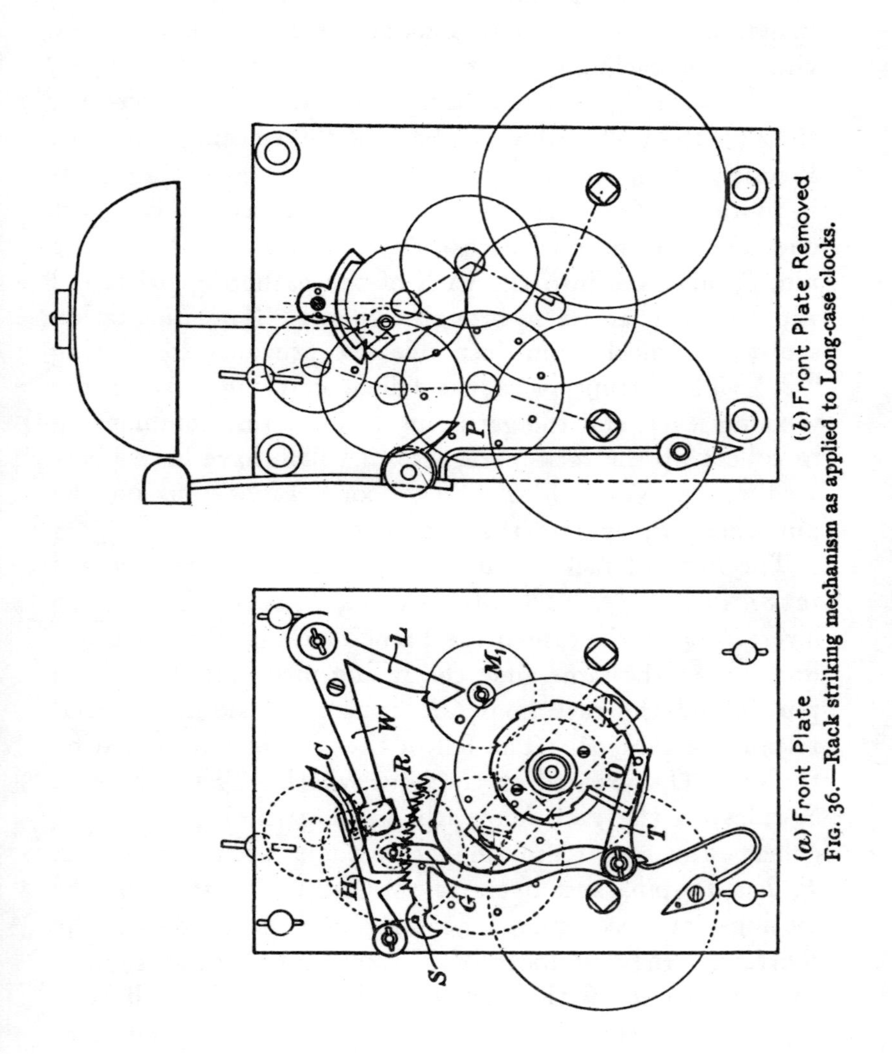

FIG. 36.—Rack striking mechanism as applied to Long-case clocks.

be struck are liberated. At the same time the gathering pallet, *G*, has been released from the stop pin, *S*, at the end of the rack and the train advances until the pin, *C*, on the warning wheel is caught on the stop projecting through the frame from the back of the warning lever, *W*. Precisely at the hour, *L* falls and the whole striking is set in motion. The gathering pallet then proceeds to collect one tooth of the rack with each revolution until the stop pin, *S*, advances into the path of the gathering pallet tail and the mechanism is brought to rest. The illustration shows the mechanism just released previous to striking "2." The lifting piece has fallen, carrying with it the warning lever, and the gathering pallet is just commencing to collect up the rack teeth, two of which have been liberated to engage with the rack hook. Eight pins on the pin wheel, *P*, operate the hammers.

The form of rack striking as applied to French clocks is shown in Fig. 37. The lifting piece, *L*, raises the warning lever, *W*, as in the case of the French locking-plate mechanism, but at the same time the lifting piece engages with a pin behind the rack hook, *H*, causing it also to rise and release the rack, *R*. The fall of the rack is determined by the snail, *O*, mounted on the hour wheel. The stop lever, *S*, rides on the rack hook arbor and gathering pallet, *G*, collects the rack teeth one at a time. The pin wheel, *P*, has 10 pins and operates the hammer just as in the locking-plate system. The half-hour is produced by shortening the first tooth of the rack and checking slightly the movement of the rack hook by fixing the half-hour pin closer to the minute-wheel centre than the hour pin, thus reducing the lift of the lifting piece. The rack, instead of dropping on to the snail, is caught by the rack hook on the second tooth, leaving only the one tooth to be gathered up and one blow of the hammer to be struck.

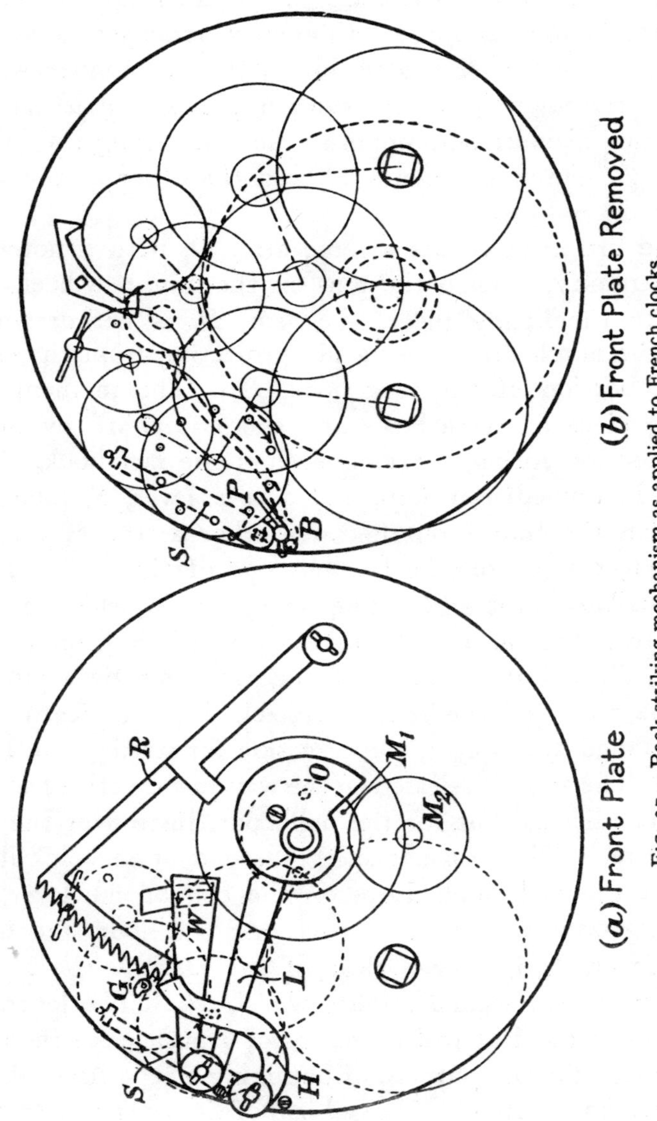

FIG. 37.—Rack striking mechanism as applied to French clocks.

Carriage Clock Striking. In French carriage clocks the mechanism is rather differently arranged, a typical example being as shown in Fig. 38. The minute wheel, M_1, corresponding to the cannon pinion, carries at the back two diametrically opposite pins for raising the lifting piece, *L*, one at the hours, *A*, and the other at the half-hours, *B*.

The lifting piece has a loose arm, *D*, with a notch at its extremity. When the pins on the minute wheel, M_1, operate the lifting piece, the arm, *D*, is withdrawn so that the notch drops over a pin projecting from the elongated portion of the rack hook, *H*. The moment the lifting piece is released the arm shoots forward by means of a tension spring, carrying with it the rack hook. The rack, *R*, immediately falls, and a stop lever, *S*, mounted between the frames on the same arbor as the rack hook, disengages a pin on the last but one of the train wheels and striking proceeds. The snail, *O*, in this case is mounted, not on the hour wheel itself, but on a star wheel, *V*, and it is advanced once an hour by a pin, *C*, on the front of the minute wheel, M_1. A feature of importance is, however, the provision for striking the half-hour. The train is released in the way just described under both conditions, but for the half-hour, instead of the tail of the rack falling on the snail, the end of the rack itself nearest the rack hook drops on the end of the lever, *F*, allowing only one tooth to pass. At the hour, the half-hour lever, *F*, is moved clear of the rack by the action of a pin on the second minute wheel, M_2. The lever, *E*, can be depressed at will by means of a push piece inserted in the case for the purpose of obtaining a repetition of the previous hour struck. The tail end of the lever just throws out the half-hour lever, *F*, and the rack hook. At one end of the horizontal part of this lever, *E*, a pin, *Q*, projects

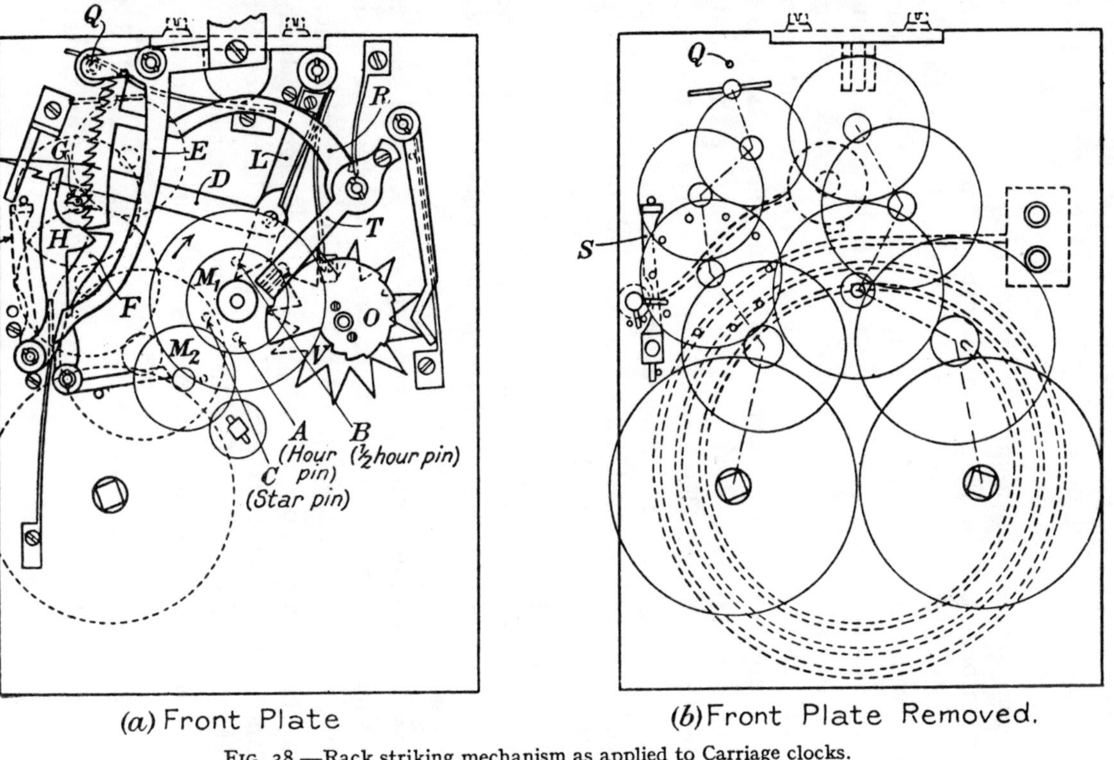

FIG. 38.—Rack striking mechanism as applied to Carriage clocks.

through the frame and catches the fly until the push piece is released, so that no striking will take place if the push piece is held down.

One other detail about this mechanism is deserving of attention. On the front and almost covering the tail, *T*, of the rack there is a thin strip of spring steel fastened by a screw at one end but free at the other. At the free end a small half-round block projects at right angles, taking the place of the pin in the ordinary form of rack striking, which drops on the snail when the rack falls. It is at times of emergency that the spring comes into play. If by any chance the striking is allowed to run down or the hands are set backwards when the tail of the rack is in the path of the snail, disaster is averted by the block mounting the deep edge of the snail, which is bevelled, on to the face. Ultimately the block clears the snail, either as a result of the continued motion of the snail or through striking off.

It should be noted that this mechanism provides for no warning, the release of the rack and stop following in quick succession before the striking.

Chiming. For the purpose of producing chimes a further train of wheels and pinions, quite distinct from the going and striking, is required. Chimes are performed on bells, gongs, tubes or rods, and the selection is usually either that known as the Westminster Chimes or the Whittington Chimes. The first is the familiar chime on four bells, said to have been arranged, in 1793, by Dr. Crotch from the opening phrase of Handel's aria, " I know that my Redeemer liveth." These chimes were designed originally for the clock at Great St. Mary's, Cambridge, and were first known as the Cambridge Chimes; later, they were also fitted in the clock above the Houses of Parliament, and the title " Westminster Chimes " then became popular. The " Whittington " is a chime of eight bells, probably deriving its name from the similarity in sound to the Bow church bells, which are reputed to have called young Dick Whittington to fame and fortune.

In most English bracket and long-case clocks the quarter-chiming train releases the hour striking. Fig. 39 shows a typical arrangement of levers, though the design may be varied slightly according to the style of case or type of chime. On the striking side it will be noticed that the rack hook, H_2, and warning lever, W_2, are placed differently from the plain striking arrangement shown in Fig. 36. The hour snail, O_2, is mounted on a star wheel advanced once an hour by a pin on the minute wheel, M_2. The quarter snail, O_1, is mounted on the minute wheel, M_1, and carries four pins, placed at intervals for raising the quarter lifting piece, L_1. The pins for operating the 1st, 2nd and 3rd quarters are equidistantly separated, but the 4th raises the lever sooner and is timed to complete the long chime and release the hour, so that the first blow comes as near as possible on the stroke of the hour. The lifting piece, L_1, carries with it the warning lever, W_1, as well as an intermediate lever, I, to raise the rack hook, H_1, and thus release the quarter rack, R_1. At the hour, the quarter rack drops on the deepest part of the snail and as this happens, the stop pin, S_1, on the end of the rack, in butting against the tail of the hour rack hook, H_2, causes the release of the hour rack, R_2. Operating in the other direction, the stop pin, S_1, on the quarter rack normally holds the hour warning lever, W_2, out of action, but when the quarter rack falls, this warning lever springs into position to check the hour train. When the chiming is nearly completed and the last tooth of the quarter rack is being gathered up, the hour warning lever, W_2, is simultaneously withdrawn, by the stop pin, S_1, and the hour striking is thus liberated.

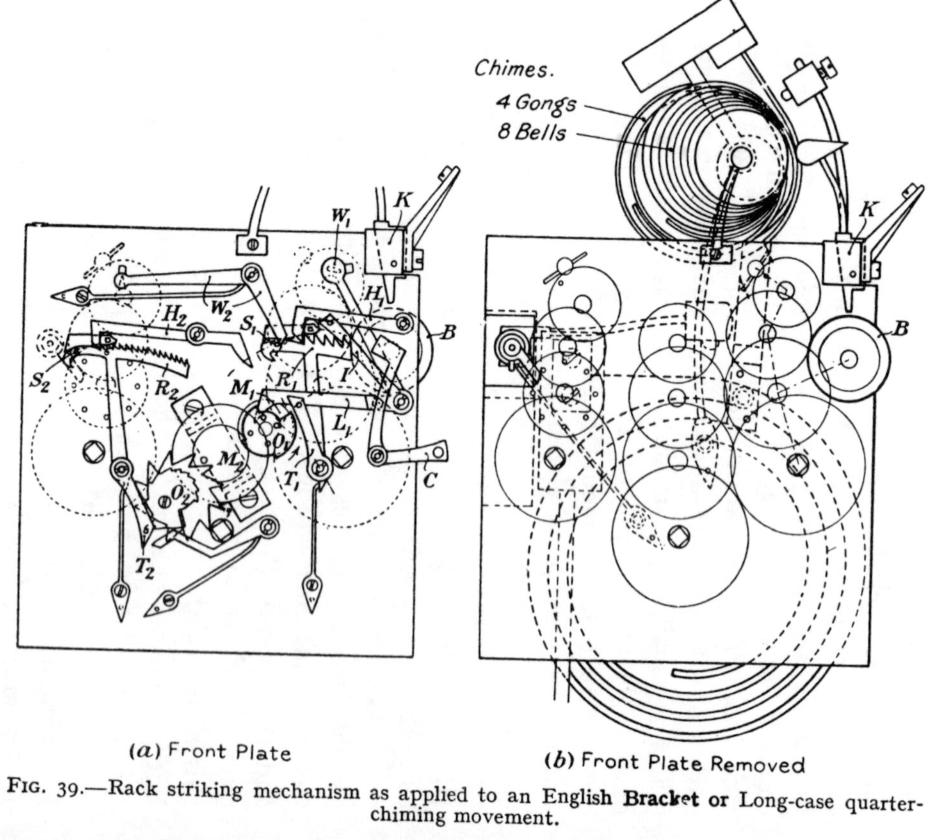

(a) Front Plate

(b) Front Plate Removed

FIG. 39.—Rack striking mechanism as applied to an English **Bracket** or Long-case quarter-chiming movement.

In the chiming train, the wheel, on the arbor of which the pin barrel, *B*, is mounted, is run off the second wheel. The second wheel corresponds to the pin wheel of the striking train. The pin-barrel arbor is pivoted with a spring endpiece at one extremity to allow a small amount of movement horizontally. Actuated at will from the front of the dial by means of a lever, *C*, this motion is arranged for the purpose of bringing the hammer tails, projecting below the hammer bracket, *K*, into the paths of a different series of pins in order to produce a change of chime. In the Whittington Chimes on eight bells, the pin barrel contains 80 pins in groups of 8, 16, 24 and 32, each separated by a space. In the Westminster Chimes there are only 40 pins, in groups of 4, 8, 12 and 16. The figure shows a stand of eight bells for the Whittington Chimes and four gongs for the Westminster, the hours being struck on a separate gong one octave deeper in tone than the tenor gong, which is the key note.

The bell stand is attached to the movement, whilst the gongs are bracketed to the case.

Many fine clocks are provided with tubular chimes arranged parallel to the back of the case. The hammer bracket is then mounted on the back plate and not between the frames.

Innumerable German chime clocks, both long case and bracket, are met with nowadays, and a most popular and effective method of chiming is adopted in these clocks. Instead of gongs or bells, rods are employed, the longest being not much more than 12 inches. They are driven or screwed into a cylindrical cast gun-metal sound box, which assists free vibration and gives resonance. As these are mass-production clocks there are many features about them which offend the eye of the artistic horologist, and many parts which would be constructed from forgings or castings in an English clock are often merely pieces of twisted wire or soft stampings. It is not a singular experience that bolder repairers, in attempting to display their prowess, find themselves hopelessly entangled in these unfortunate pieces of wire and some have been known to admit defeat and give up the job in despair.

Turret Striking and Chiming. Plate V (Fig. 40) shows a typical example of a modern quarter-chiming turret movement for four bells (Westminster Chimes). The going train is arranged in the centre of the frame terminating in a gravity escapement. The minute-hand drive is taken off the main wheel, *M*, through a vertical connecting rod and the bevel work, *B*. The latter distributes the drive to four sets of motion works situated behind their respective dials. On the left of the going train is the striking train with its locking plate, P_1, and lifting piece, L_1, which is raised by a cam on the hour arbor. On the right is the chiming train, showing distinctly the cam wheel, *C*. Five hammers are provided, an extra one being needed to permit the tenor bell being struck twice in succession when it so occurs during the third and fourth quarters. The locking plate, P_2, determines the appropriate length of the quarter similar to the one for the hours. The lifting piece, L_2, for releasing the quarters is raised by the same cam as the hour lifting piece, L_1. The cam has four arms, the one operating the hour lifting piece, which is mounted in a different plane, being wider than the other three. The striking and chiming trains are not connected in any way, but it is arranged for the fourth quarter to complete early enough to allow the first blow of the hour to be on time. At other quarters, the first blow of the quarter should be correct. Plate VI shows another view of the same movement.

PLATE V (Fig. 40).—A modern turret chiming movement. This clock is fitted with a double three-legged gravity escapement and is arranged for giving Westminster Chimes.

[*To face page 102.*

PLATE VI.—A different view of the turret chiming movement shown in Plate V. The five chiming hammer levers can be seen distinctly, the hour hammer lever being at the opposite end.

[*To face page 103.*

CHAPTER IX

CALENDARS, ALARUMS AND SUPPLEMENTARY DEVICES

Calendars. Numerous and varied are the specimens of supplementary devices embellishing older clocks. Perhaps the most general features met with are the date circle and lunar phases frequently fitted to grandfather clocks. A wheel gearing with the hour wheel and mounted on a stud rotates once in 24 hours. This wheel carries a pawl projecting from the face, which, on the completion of rotation, advances one tooth of an annular ratchet. This ratchet is otherwise just a circle forming part of the dial itself and, having a smooth periphery, it is free to slip round on four guiding pulleys, which also hold it in position. It has 31 ratchet teeth and on the front the numbers are engraved corresponding to the days of the month and, one at a time, they are to be seen through a little square aperture in the dial. At the end of a short month the remaining numbers have to be pushed round by hand.

The lunar phases are displayed in the semicircle at the top of an "arched" dial. The moon rotates round the earth in about 29½ days, so that to overcome the difficulty of recording the half-day the mechanism is always arranged with two representations of the moon appearing alternately. The "moons" are painted "full" opposite one another on a toothed disc, and the background is painted to represent a starry firmament. The disc has 59 teeth, which, together with the half not to be exhibited, is appropriately screened by the dial. Following the "new moon" a segment commences to appear on the left-hand side of the arch and, being advanced day by day similarly to the date circle, travels gradually over the whole aperture and disappears on the right-hand side.

Sometimes numbers are given on the top of the arch to show "high water" at a particular place, London, Liverpool, etc., but these serve only as an approximation.

The date circle often appears also in bracket clocks, but comparatively rarely does one meet with a device to show the months in a clock. To cite, however, an exception, Brocot introduced a perpetual calendar mechanism * as a supplement to some of his fine French clocks. He planned these in slightly different ways, using a double case, so that the ordinary movement and calendar movement with their respective dials appeared independently side by side. Once a day, the lever connecting the two would be operated by the going train and the day of the week wheel would be thus advanced. The unequal lengths of the months was adjusted by means of a locking plate with a series of slots. This was advanced once a month and performed a complete revolution in four years so that leap year might be included. According to the depth of the slot, a lever would alter the setting of the day lever in such a way as to cause it to advance 2, 3 or 4 days in one operation as occasion required.

Carriage Clock Alarums. An alarum is an appendage sometimes met with in good carriage clocks, the mechanism of which is shown in Fig. 41. The alarum wheel, *A*, is identical with the hour wheel, *H*, by which it is operated through an intermediate wheel, *I*, also the same size and number. A notched disc, *D*, is attached to an arbor carrying a hand which indicates, on a small circle at the foot of the dial, the hour at which the alarum is to be set. The alarum wheel rides freely on this arbor, but is sprung

* A description of one form of these is to be found in Saunier's *Treatise on Modern Horology*.

from the back so that a pin or thin block projecting from the collet on which the wheel is mounted impinges continuously against the notched disc. At the appointed time, the block drops into the notch, which is undercut to ensure definite action. This causes the disengagement of two levers, *L*, L_1, one of which, planted horizontally and behind the wheel, follows the movement of the block,

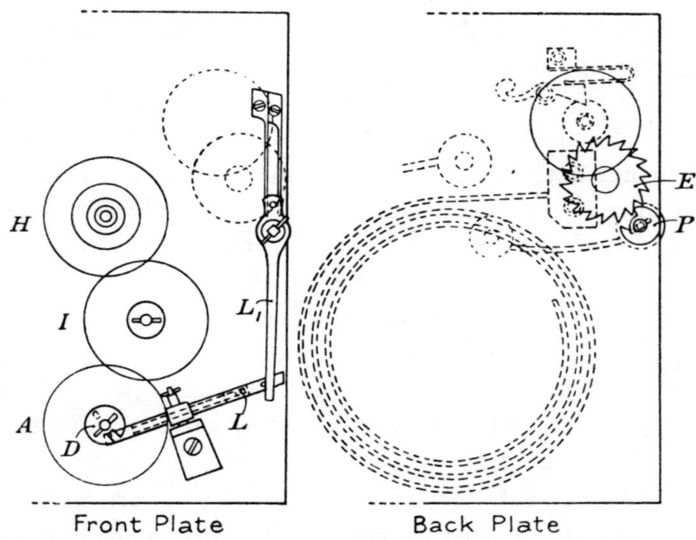

FIG. 41.—Showing an Alarum mechanism as applied to a Carriage clock.

whilst the other planted vertically communicates with the alarum. As soon as the vertical lever is free the alarum goes off. The alarum itself consists of a recoil escapement, *E*, worked by a small mainspring and planted in the frame. An arm attached to the pallets, *P*, carries a hammer which strikes the hour gong or, sometimes, a specially provided small bell. When the alarum has taken place and the alarum wheel continues its motion, the block rides up a slope at the back of the notch and sets the horizontal lever, *L*, again in the path of the vertical lever, L_1.

CHAPTER X

ELECTRICAL CLOCKS

THE designs and devices connected with the evolution of electrical horology are so very numerous that it is intended to deal in this chapter with the principles underlying only those mechanisms which have latterly come into greatest prominence both scientifically and commercially. It cannot be disputed that electricity as applied to clockwork now forms one of the most important branches of horological science, and still offers scope for development along lines of simplicity in mechanical construction.

Experimental work on the subject dates back to the commencement of the Victorian era, though, strangely enough, present attainments can really only be said to be due to the advancement made during the last forty to fifty years. During the 'forties several devices concerned with the application of electricity to timepieces were patented by Alexander Bain, who will always be remembered as one of the pioneers. It seems probable, however, that this scientist, and others who devoted themselves to the question of maintaining the motion of a pendulum by electrical means, had at heart the solution of the problem of perpetual motion quite as much as the advancement of horology. The very system adopted by Bain for obtaining electrical energy seems to point to this. He generated the necessary current by means of a zinc and copper plate buried in moist earth. If, therefore, this conjecture is at all accurate it may reasonably account for the long period of sterility in the advancement of electrical horology.

One of Bain's clocks may still be seen (and usually working) in the Science Museum at South Kensington. The pendulum of this clock carries a coil in place of the bob, which moves in the field between two fixed magnets with North poles adjacent. One of these attracts the coil and the other repels, and the pendulum is arranged to receive impulse only when passing to the right. This is effected by means of a " make and break " which consists of a slide moved to and fro by a pin half-way down the pendulum rod. The ends of the slide turn down at right angles and pass along grooves cut diametrically across two gold and agate pieces, one at each end. When the points rest on the sections of gold, contact is made; but when the points are on the agate, the circuit is interrupted. Should the pendulum receive more impulse than desirable, so that the amplitude of the arc becomes excessive, then the slide is carried farther towards the left and off the gold piece, with the result that no contact is made and the pendulum receives no impulse on the following beat. The clock itself is of a usual type, but has no weight or other motive power, since the pendulum receives impulse independently. The escapement, therefore, is formed in such a way as to be driven by the pendulum and the order of the going is thus reversed. Reference is made later to a clock introduced by C. Bentley which follows very closely the principle first devised by Bain.

Generally speaking, it may be preferable to think of electrical clocks as comprising two main groups, namely, Primaries and Secondaries; or Time Transmitters and Time Recorders. The clocks in each group, however, may differ considerably in principle and application.

PRIMARIES OR TIME TRANSMITTERS

The primary group includes the mechanisms which actually measure periods of time, not necessarily recording, whilst those of the secondary group are mechanisms electrically synchronised periodically at short or long intervals by the primary, that is to say, they are simply governed by the behaviour of the primary; what the one dictates the other records.

Again those of the primary group may be either: (1) a clock which is electrically wound at regular intervals; (2) a mechanism in which the pendulum receives impulse by electro-magnetic means; or (3) a super-mechanical clock, such as a regulator, provided with an attachment for distributing signals or closing a circuit at specific intervals by means of electrical contacts.

The greatest possible accuracy of timekeeping is, of course, expected of these machines, especially if, for the control or synchronisation of a system of clocks, they are employed as standard "Master" timepieces.

Electrically-wound Primaries. Of the three categories just mentioned, the first has probably been developed to greater advantage than the other two, chiefly on account of the fact that by such means it is possible to ensure a maximum amount of freedom to the pendulum. The most usual way of applying this principle is to reset a weighted arm or rewind a flat spring by means of the attraction, at a given moment, of an armature to an electro-magnet. Rewinding a weight or spring by automatically running an electric motor is, however, another method which may be adopted.

The idea of providing impulse for the pendulum by releasing a gravity lever was patented as far back as 1849 by Charles Shepherd, and it is interesting to record that such a clock of his construction was for many years in use at the Royal Observatory, Greenwich.

It was not, however, until 1895 that electrical horology commenced to advance commercially. In that year and in 1897 the "Synchronome" Co. (then Bowell and Hope-Jones) patented their earliest forms of gravity impulse transmitter. Numerous developments on similar lines came along in the decade which followed, notably, the Standard Time Co.'s transmitters and the "Pulsynetic," whilst in the "Lowne" primary a propulsion spring impulse was adopted. The recently introduced "Princeps" primary is somewhat different in principle from those just mentioned, but in this instance also impulse is provided by a spring. All these systems, excepting the "Princeps," employ a count wheel, which is operated with the swing of the pendulum and determines the equal intervals necessary for applying impulse. Generally these intervals are of 30 seconds' duration. The chief feature to notice is, however, that by employing a mechanically liberated impulse of constant force and period the pendulum is rendered free of any disturbing influence due to fluctuation of either voltage or supply of electric current.

The "Synchronome" Primary. Fig. 42 illustrates diagrammatically the system patented by the Synchronome Co. The pendulum rod, *P*, carries a projection, *M*, the extremity of which forms an impulse plane for the roller, *R*, pivoted on the bell-crank lever, *L*. Extending from the projection, *M*, a light pawl with a hooked end passes over the ratchet-shaped teeth of the count wheel, *W*, as the pendulum swings to the left, but gathers up a tooth on its return. The wheel has 15 teeth and is advanced one tooth with every double beat, back action being checked by the jumper, *J*, so that a complete rotation is made every half minute. A finger, *F*, rotates with the wheel and once in every revolution disengages the catch, *C*, thus releasing the lever, *L*. The roller, *R*, drops on the pendulum projection, *M*, and in descending the inclined plane a gravity impulse is provided just as the pendulum is crossing the line of centres. At the end of its fall, the lever, *L*, makes contact with the armature, *A*, which, being thus attracted to the electro-magnet, *E*, throws back the lever, *L*, over the catch, *C*. Contact is broken and the armature falls away, leaving the lever, with its impulse roller, locked in position.

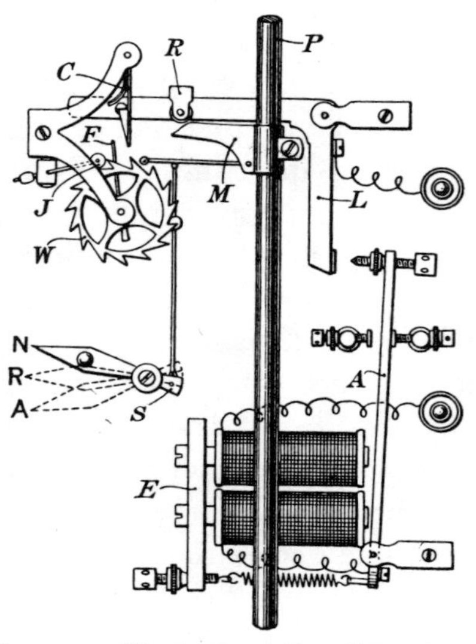

Fig. 42.—Illustrating the "Synchronome" design of electrical primary.

Simultaneously with the making of contact for resetting the gravity impulse lever, the same contact is used for closing another circuit, so that a system of secondary clocks may be synchronised. Should these require alteration or adjustment for time, the operation of correcting them is carried out from the primary or master clock by means of the control lever, *S*. This is capable of being moved by hand from its normal position, **N**, as shown, to two other positions, raising at the same time the pawl extending from the projection, *M*. At position, **R**, the pawl is raised free of the count wheel and progress of the secondaries is retarded for 2, 4, 6, 8, 10 seconds and so on as required. At position **A**, however, the pawl is raised so high that with every alternate beat the catch, *C*, is disengaged and the impulse unlocked, with the result that contact is made every 2 seconds instead of every 30 seconds, and half-minute advances of the secondaries are made in rapid succession. Provision is also made in this system for giving warning of failing current supply by sounding a bell or lighting a lamp. If the supply becomes too weak to reset the gravity impulse lever, the magnet is assisted by the pendulum itself raising the impulse roller during the swing to the left and closing an auxiliary circuit.

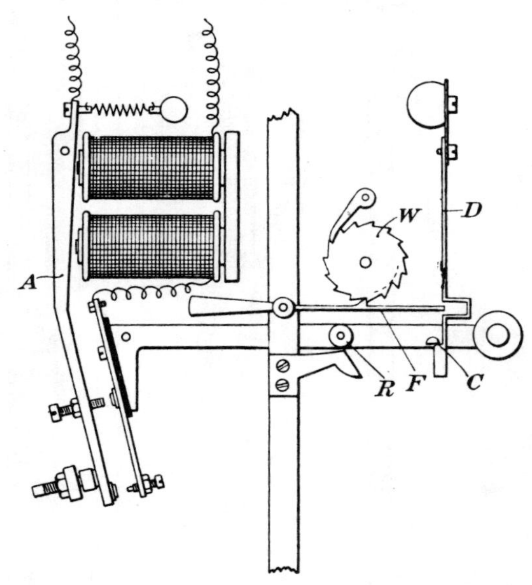

FIG. 43.—Illustrating the "Pulsynetic" primary.

The "Pulsynetic" Primary. The "Pulsynetic" (formerly known as the Parsons and Ball) system shown in Fig. 43 is very similar to the one just described. In this case, however, it will be seen that the count wheel is pushed instead of pulled whilst the pendulum is proceeding to the right. One of the ratchet teeth on the count wheel, *W*, is cut deeper than the rest and when the arm, *F*, drops into it, the detent, *D*, which normally is missed on account of the recess, becomes deflected and releases the catch, *C*, to allow the gravity arm and impulse roller, *R*, to fall until reset by making contact with the armature lever, *A*.

The "Standard Time Co.'s Primary." Fig. 44 illustrates the type of primary introduced into the Standard

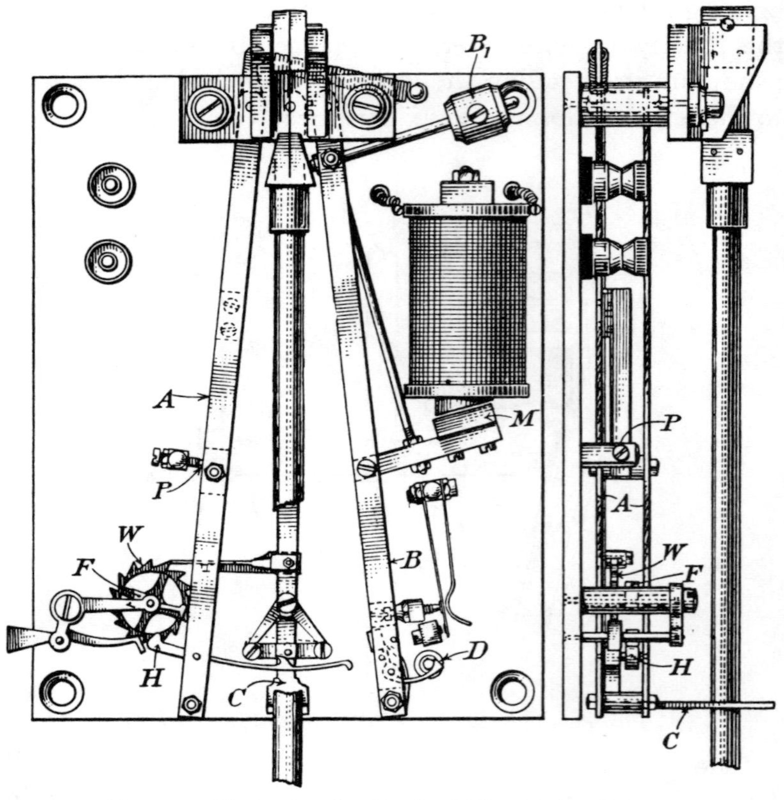

Fig. 44.—Illustrating the Standard Time Co.'s primary.

Time Co.'s system. Here the pendulum receives its impulse through a crutch, *C*, after the principle of those used in mechanical clocks. The crutch is pivoted near the suspension, and on either side there are also two long pivoted levers, *A* and *B*. The lever, *B*, with its weighted end, B_1, is actually the gravity arm providing impulse through the crutch to the pendulum, but it is held back by a catch and trigger, *D*, until at a given instant determined by the count wheel, *W*, it is released. The count wheel is propelled at alternate beats by a pawl mounted on the crutch. The lever, *A*, in its normal position makes contact at *P*, and near the bottom a light horizontal lever, *H*, is pivoted, the short end of which is bent at right angles towards the count wheel, *W*, whilst the long end projects towards the crutch. On the completion of each rotation, a finger, *F*, travelling round with the count wheel depresses the tail of the lever, *H*. A hook on the front part of the lever, *H*, is by this means raised into the path of a pin, projecting from the crutch, *C*. As the pendulum swinging to the right crosses the line of centres the pin on the crutch is lifted over the hook on the lever, *H*, and the pendulum thus carries with it the entire lever, *A*, which breaks contact at *P*. A projecting portion of the lever, *H*, beyond the hook finally releases the trigger and catch, *D*, causing the gravity arm, *B*, to exert its impulse on the return swing to the left. Just after the crutch has again crossed the line of centres, the lever, *A*, makes contact at *P*, the electro-magnet is energised and the gravity arm, *B*, is reset through the armature, *M*. As the crutch continues to the left the hook drops off the pin, leaving the tail again in the path of the count wheel finger.

The "Lowne" Primary. The three systems just described are all gravity impulse clocks, but the Lowne primary, shown in Fig. 45, furnishes an example of a spring impulse clock and, as in the last case, a crutch is employed as the medium for delivering the impulse. A count wheel, *W*, is again used and advanced by a pawl on the crutch, *C*, during the swing of the pendulum to the left. Two spring contacts, P_1 and P_2, are attached to the crutch at right angles, so that the points have a slightly circular movement up and down. Frictional contact is, however, actually made on their motion being arrested by the end of a finger rotating with the count wheel. This occurs with every half revolution of the count wheel when the pendulum is travelling to the left and, as soon as the circuit is closed, the armature, *A*, is attracted, contact at P_3 is broken and, at the same time, the pin on the pivoted lever, *L*, rides over

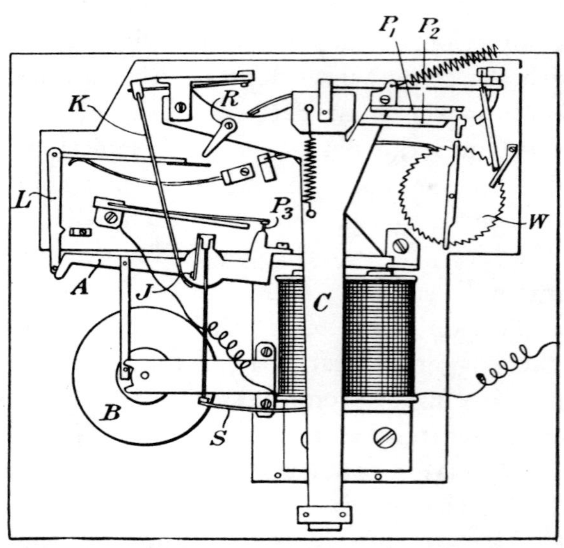

FIG. 45.—Illustrating the " Lowne " primary.

the tail of the armature, *A*. The armature is thus locked, the impulse lever, *K*, is set into the catch, *J*, and the impulse spring, *S*, compressed. As the crutch completes its return vibration, the projection, *R*, depresses the lever, *L*, which in turn releases the armature, *A*. The impulse spring, *S*, then raises the armature and with it the impulse rod, *K*, as well as the crutch through which the actual impulse is transmitted to the pendulum. The adjustable disc, *B*, provides a means of controlling and regulating the speed of the armature.

The "Princeps" Primary. One of the latest introductions into the field of electrical horology is the "Princeps" clock, the pendulum of which also receives spring impulses. A noticeable deviation from the usual custom in the construction of these clocks is that impulse is given with alternate beats and no count wheel is employed. The electrical part of the mechanism consists of a polarised electro-magnet arrangement known as the reverser, which is shown diagrammatically in Fig. 46. An armature in the

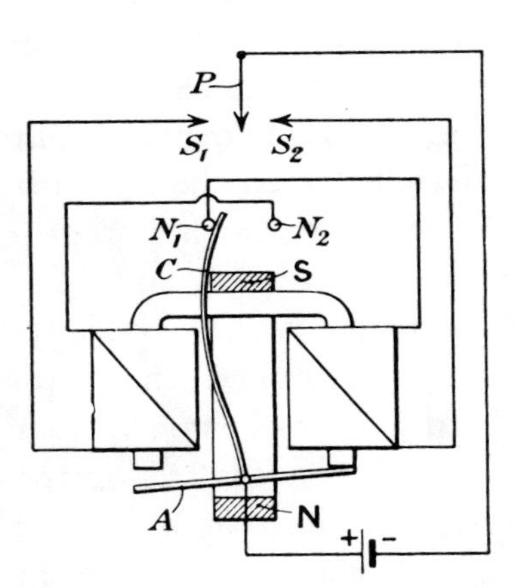

FIG. 46.—Showing the Reverser in the "Princeps" primary.

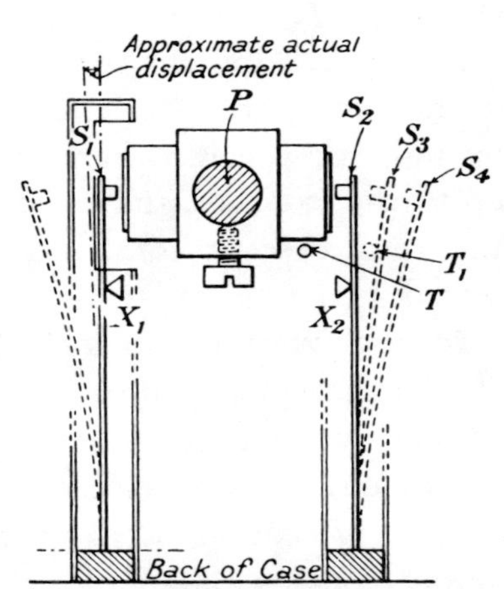

FIG. 47.—Illustrating the "Princeps" method of giving impulse.

form of a rocker arm, A, is attracted first to one solenoid and then to the other, according to the direction of the flow of current, this being dictated by the movement of the pendulum, P, between two contact springs, S_1, S_2. With the change over of the armature, A, the contact arm, C, also passes between the points, N_1, N_2, and by this means closes a secondary circuit for synchronising purposes. Fig. 47 shows diagrammatically the pendulum viewed from above and the method whereby impulse is given. The pendulum, P, vibrates between the two contact springs,

S_1, S_2, each of which is protected by a glass tube and provided with a stop, X_1, X_2. On one side there is an auxiliary stop, T, which is removed to the position T_1, with the attraction of the rocker, A (Fig. 46), to one of the solenoids. As the pendulum swings to the right, instead of meeting the contact spring, S_2, at rest on the stop, X_2, it has to proceed to the position shown at S_3, the spring having been already advanced by the motion of the stop from T to T_1. The pendulum then makes contact, the stop, T_1, retires because the rocker, A (Fig. 46), has changed over and the spring is carried on by the pendulum to the end of the beat at S_4. On the return to the left, the spring, acting against the pendulum, remains in contact right up to the stop, X_2, and the difference between the total lift and total descent constitutes the impulse which is represented in the figure by the separation between S_3 and S_2.

Motor-wound Clocks. In addition to the varieties of electrically-wound clocks already described there are also many instances where an ordinary mechanical clock is provided with an electrical winding gear for raising the weight or coiling the mainspring. The winding gear simply consists of an electric motor which is started at regular intervals by means of contact switches in the clock mechanism itself. In the case of a weight clock, the length of fall available for the weight would determine the suitable period between the successive re-windings. The method employed in spring clocks wound in a similar manner may more conveniently take the form of a worm drive on the shaft of the motor gearing into a wheel mounted on the barrel arbor. In both types, care has to be taken to ensure that the motor is switched off in time to prevent overwinding, so that everything really depends on the design of a suitable device for starting the motor and bringing it to rest when its work has been performed.

Steuart's Continuous-motion Clock. The latest development in motor-wound clocks is the ingenious mechanism known as the Steuart continuous-motion clock, the principle of which is illustrated in Fig. 48. Behind the pendulum two gravity arms, *A*, *B*, are pivoted on the line of centres and at the bending line of the suspension spring. The arm, *B*, carries the disc, *W*, mounted just below the centre, which acts as the gravity-weight, keeping

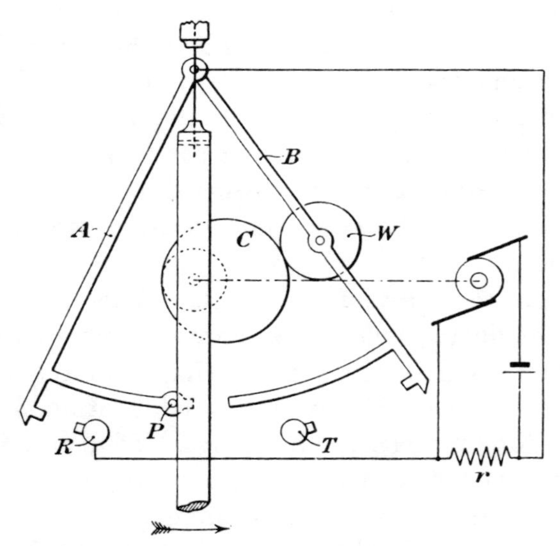

FIG. 48.—Steuart's continuous motion clock.

the disc, *W*, in contact with the eccentric cam, *C*. The latter rotates on the arbor of a wheel pivoted behind the pendulum, on the line of centres, and is driven by a pinion on the shaft of an electric motor. The motor is in continuous motion, though the supply of current is intermittent. Impulse is given to the pendulum whilst swinging to the right through contact with a pin, *P*, which projects from the gravity arm, *A*. When the extremity of the gravity arm, *A*, reaches the adjustable stop and contact,

R, the direct current through the motor which maintains the rotation of the cam, *C*, is short-circuited through the resistance, *r*. As the pendulum proceeds still farther to the right, the cam, *C*, gradually presents its shortest radius to the wheel, *W*, and so lowers the arm, *B*, until its extremity reaches the stop, *T*. In doing this the arm, *B*, also raises the impulse arm, *A*, to its set position, breaking contact at *R* in the process. When the pendulum has arrived back at the extreme left, the cam, *C*, has already commenced to move the arm, *B*, away to the right, preceding the pendulum and leaving the arm, *A*, free to supply impulse to the pendulum as before. Between the periods of contact the motor is kept running partly by kinetic energy, but also by a reduced current passing through the resistance, *r*, which serves at the same time to eliminate sparking at the contacts.

Although the motion of the motor is not perfectly regular, the fluctuations of speed are very small and do not prevent sub-divisions of a second being accurately observed. Whenever the necessity for measured continuous motion arises, as, for example, in the rotation of drums for chart recording or for driving astronomical telescopes, this mechanism may be utilised to advantage.

Before passing to other types of electrical clocks there are certain features of contrast in connection with the foregoing systems which must not be left unnoticed, notably in the manner in which the respective pendulums receive impulse. Considering the systems individually, it will be observed that the "Synchronome," "Pulsynetic." "Standard Time" and "Lowne" all employ a count wheel which has to be operated by the pendulum, and impulse is given intermittently, generally at half-minute intervals. In the case of the first two, the pendulum receives impulse near the line of centres, where, aided by gravity, it has achieved its maximum velocity; whilst, in the other two instances, impulse is given at the commencement of the beat, ceasing soon after crossing the line of centres. The "Steuart" clock is similar in this respect, but differs in three other important aspects—the impulse is continued farther across the line of centres, delivered as frequently as at alternate beats, or even every beat, and the pendulum is entirely relieved of the work of unlocking. The "Princeps," however, stands in contrast to the rest, first, inasmuch as impulse is delivered only during a short period before the pendulum reaches the line of centres, and, secondly, that the pendulum has to raise side contact springs at every beat. The friction introduced by these contact springs suggests a similarity to the unlocking friction of a mechanical escapement, but whereas the latter presents point contact, the springs of the former are affected by frictional resistance. This condition may, however, be regarded as constant, particularly as it occurs with each vibration.

As emphasised in Chapter III, on Pendulums, the ideal for which one strives in order to ensure vibrations as nearly perfect as possible, is to remove every particle of obstruction or interference with the beat. The series of mechanisms just described, with the exception of the motor-wound ordinary clocks, do approach very closely that free pendulum ideal, provided that the essential feature of good workmanship is maintained.

Electro-magnetic Impulse Clocks. Clocks of the primary group following the second category, namely, those in which the pendulum receives impulse through an electro-magnetic agency, are by no means uncommon. Bain's clocks, one of which was described in the early part of this chapter, were of this type. An objectionable feature is that the pendulum is open to the effect of any disturbing influence due to fluctuation in the electric current, with consequent variations of arc. Excepting in the case of the "Bentley" clock, where this difficulty is overcome, these fluctuations are most detrimental in instances where the greatest precision is essential. Nevertheless, clocks of this class have been constructed in large numbers, though not often as "Master" timepieces.

The Bentley Clock. Following very closely upon the principle of the Bain clock, inasmuch as it is maintained by a voltaic current through zinc and retort carbon plates buried in the earth, the Bentley clock was introduced in

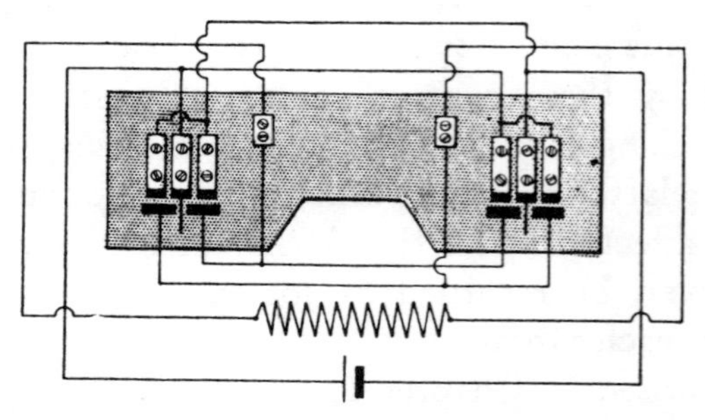

FIG. 49.—The "Bentley" clock. Showing the arrangement of the contacts across which the "make and break" carriage travels as determined by the arc described by the pendulum.

1912. Instead of Bain's "make and break" slide, a bar is used with a little wheel at each end. The pendulum carries a coil above the bob and there are two fixed magnets with adjacent north poles intercepting the field of the coil as before. The chief feature is, however, that the arc of vibration is more closely under control than in Bain's clock. The wheels at the extremities of the bar or carriage travel upon groups of contacts (Fig. 49), which are so arranged that the direction of the current is reversed as soon as a certain arc has been completed; if, however, the arc becomes excessive and overruns the point, the carriage wheels reach contacts which maintain the original direction of the current and retard the downward acceleration due to gravity. This clock, therefore, is rather an exception to the rule of its kind, because by means of the contact device the arcs are kept as near constant as possible.

Hipp Trailer. In this category of electro-magnetic impulse clocks the principle which has been developed and improvised upon to a very considerable extent was originated by Dr. Hipp about the year 1850. This system, one application of which is shown in Fig. 50, has the advantage of being comparatively economical in current consumption and extremely simple in construction. At the foot of the pendulum, below the rating nut, a soft iron bar forming an armature is attached. Under this, fixed to the case and slightly to one side of the line of centres, is an electro-magnet. Half-way down the pendulum rod a small pivoted flap is arranged to trail with the vibrations to and fro over a notched block which projects above the upper arm of a pair of contact levers screwed to a bracket on the back of the case. As the pendulum vibrates, the flap, or trailer, passes over the block so long as the arc permits. When the energy of the pendulum weakens and the arc falls off, the trailer fails to pass the notches in the block, with the result that the lever is depressed, contact made and the circuit through the magnet closed. The armature on the pendulum is at once attracted, the vibrations re-energised and the trailer again passes the block.

Impulse is thus given on the downward vibration, the intensity depending entirely on the force of the current. The vibrations will be vigorous at first, the maximum arcs wide and the intervals between the impulses great, but as the supply of current degenerates the arcs will narrow down and impulse will be sought more frequently.

Féry Principle. Another system, on rather different lines, was developed by Ch. Féry, utilising the effect of

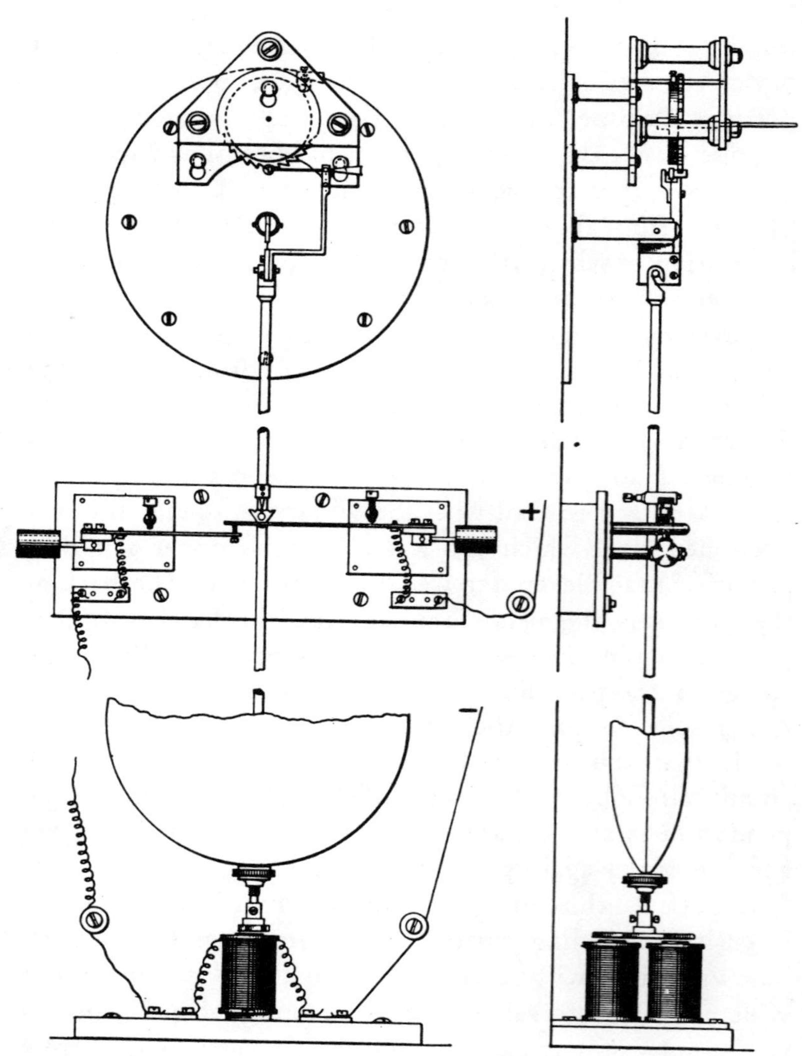

Fig. 50.—Showing an application of the "Hipp trailer" principle; from a model constructed by Mr. W. Beckmann.

Foucault currents. The eminent French physicist, Jean Foucault (1819–68), discovered that if a metallic mass lay in a magnetic field of varying intensity, electrical currents were induced in that mass. In Féry's application of this phenomenon, two pendulums are used, having the same time of vibration, and suspended in line with each other. At the bottom of one, P_2 (Fig. 51), is a copper ring, C, to admit the extremity of an arm of a horse-shoe magnet,

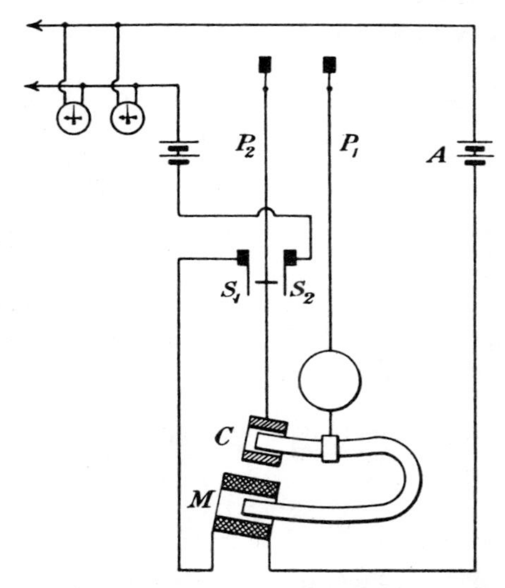

FIG. 51.—Illustrating the Féry principle of impelling a pendulum.

which is attached horizontally below the bob of the second pendulum, P_1, as shown. The other arm of the horse-shoe magnet passes in and out of a solenoid, M, as the pendulum, P_1, vibrates. During the vibration of the pendulum, P_2, contact is made alternately with the springs, S_1, S_2. Contact with S_2 closes the circuit through the secondaries, whilst contact with S_1 closes the current from the battery, A, through the solenoid, M, which attracts the one arm of the magnet, the other entering the copper ring on the pendulum,

P_2. Foucault currents are thereby induced and react on the arm attracted to the solenoid. Thus, by successive reactions, the pendulum P_1 maintains the vibrations of the pendulum P_2, and this alternately closing the different circuits, the one energising the solenoid, *M*, the other distributing signals or impulses to the secondaries, keeps the system in continuous operation.

Moulin-Favre-Bulle Clock. Although not used as a transmitter, an interesting example of an electro-magnetic impulse clock which has recently come into commercial

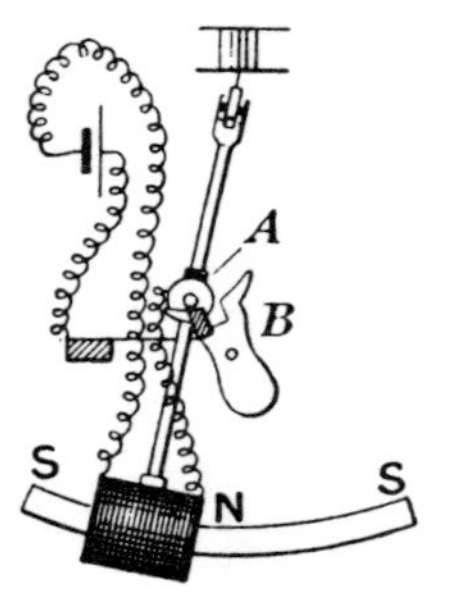

FIG. 52.—The Moulin-Favre-Bulle clock. Showing the system of electro-magnetic impulse.

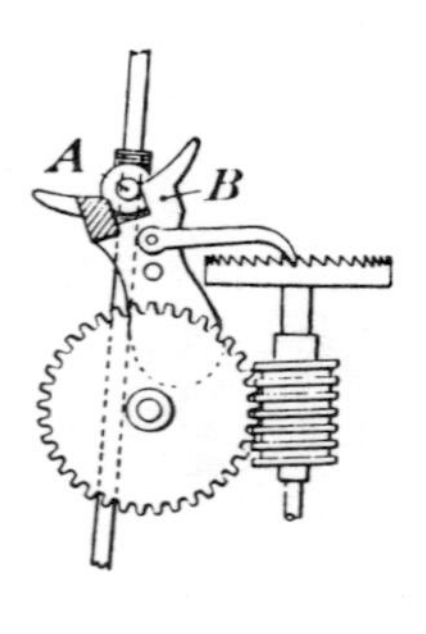

FIG. 53.—Illustrating the counting mechanism of the Moulin-Favre-Bulle clock.

prominence, principally as a domestic timepiece, is known as the "Bulle." The bob of the pendulum (Fig. 52) takes the form of a hollow coil through which is passed a curved permanent magnet, magnetised so that at both extremities there is South polarity whilst the North is concentrated about the centre. The coil travels along this magnet as the pendulum vibrates, and at the same time a pin, *A*, on the upper part of the rod oscillates with each beat a pivoted contact-piece, *B*, a combination closely resembling the ruby pin and forked part of the lever of a watch lever escapement. One of the forks is insulated so that contact is made only on alternate vibrations. The arrangement of closing the circuit through the coil, combined with the effect of the bar magnet with its central north pole, serves a dual purpose. Not only does it provide the impulse necessary, but the external supply of current from a battery, to excite the coil, is automatically diminished by reason of the electro-motive force generated in the coil by the motion of the latter with regard to the magnet. Thus the clock is maintained on a very low current consumption, whilst the vibrations have the advantage of receiving impulse regularly, which renders them more isochronous. Fig. 53 shows the way in which the contact-piece propels a horizontal count wheel at alternate beats in order to advance the hands.

Mechanical Primaries Transmitting Electrically. In this third category of primaries which simply consist of mechanical precision clocks it is the mode of electrical transmission which here comes under review. The attachment varies according to the purpose for which the signals may be required, but generally the given instant for distributing the signal is denoted by a count wheel mounted on the escape pinion arbor. The wheel is cut with V-shaped teeth placed at requisite intervals, so that, as the wheel rotates, the teeth intercept a contact arm. A regulator may be arranged in this way to send out remote sound signals in seconds with, perhaps, a gap at the 60th or at the 59th and 1st, or in numerous other ways for the benefit of an observer. In some instances, the contact may occur only at the 60th second and be used to synchronise automatically a distant secondary clock. The contact device itself may consist of light spring arms attached to an insulated bracket or, better still, the arms may be pivoted and delicately counterpoised by small adjustable gravity weights.

"**Magneta**" **Primary.** There is, however, another

form of mechanical primary embodied in the " Magneta Time System " which possesses certain unique features in the method of controlling secondary clocks and signal transmission generally. Some of these primaries are weight-driven and some spring-driven, but in either case the motive power, besides keeping the clock going, is used to rotate an iron cylinder within a coil of wire, at one-minute intervals, through a quarter of a turn at a time. The motion of the iron cylinder thus generates an alternating magneto-electric current through the coil sufficient to actuate a large number of dials or other signalling devices which may be included in the circuit. An important point to notice is that the circuit is always closed, so that faulty or corroded contacts are completely eliminated, as well as the need for batteries which would both require periodic attention.

SECONDARIES OR TIME RECORDERS

As stated earlier in this chapter, the secondary group embraces mechanisms which are constructed to record, by periodical adjustment, the performance of a primary. Some of these mechanisms may, in certain instances, take the form of a time-measuring instrument, either mechanical or electrical, which is synchronised at specific intervals by the primary. Perhaps the most important example of a secondary is, however, a simple dial mechanism, forming a unit in what may be a large system of recording time-pieces all controlled by one primary. Sound signals, as electric bells, or even wireless time signals, as well as rotating drums and charts, may all be secondaries, but a description of these is hardly necessary here.

For indoor systems, the methods of constructing dial-work secondaries vary but little, though for turret clocks, where large hands are exposed to the elements, difficulties arise which demand a more robust mechanism.

"Synchronome" Secondary. Fig. 54 illustrates the Synchronome system for ordinary requirements and embodies the principle upon which most indoor dial secondaries are founded. It is a simple device consisting of the usual type of clock motion work, but the centre arbor, on

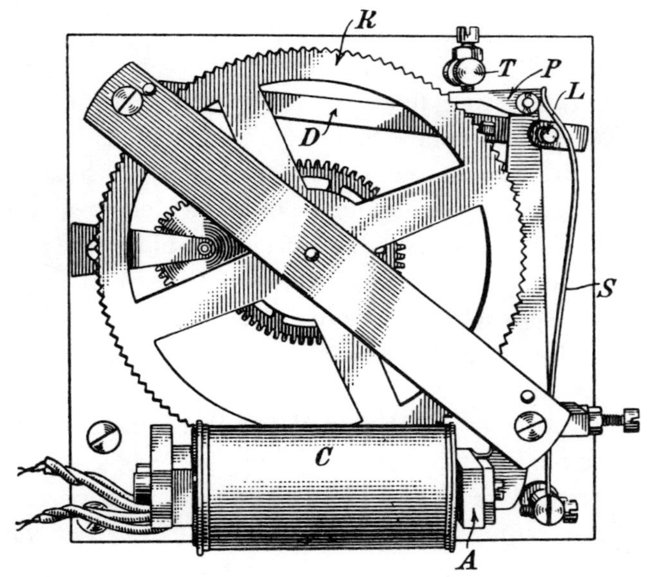

FIG. 54.—"Synchronome" secondary or dial mechanism.

which the minute hand is mounted, carries a large ratchet wheel of 120 teeth, *R*. This wheel is thus advanced at half-minute intervals, relying on the primary clock to excite the coil, *C*, when necessary and attract the pivoted armature, *A*, on which is mounted at its opposite extremity a propulsion click, *P*. Every time the circuit with the primary clock is broken the armature, *A*, falls away from the coil, aided by the spring, *S*, and the propulsion click, *P*, advances one tooth of the ratchet. The stop, *T*, controls the advance of the wheel and locks it to prevent any extraneous movement of the hands. The gravity lever or detent, *D*, which carries a pallet to drop into the wheel teeth, obviates the possibility of the wheel turning backwards when the propulsion click, *P*, is withdrawn to advance the next tooth. When the armature is attracted the propulsion click lever is brought sharply up to the stop, *L*, and the recess in the lever at that point embracing the stop temporarily locks the detent, *D*.

"Silectock" Secondary. A form of secondary clock possessing interesting features, principally that it operates quite silently, is the "Silectock." There is the usual motion work, but the centre arbor carries a wheel driven by a pinion, mounted on the arbor of which is a double-stepped snail, the steps being spaced equidistantly. The snail forms an armature free to revolve between four pole pieces—two opposite ones being the poles of a permanent horseshoe magnet and the other two those of an electro-magnet. Whilst at rest, the permanent magnet fixes the projecting portion of the snail-armature, but when the circuit is closed by the primary and the electro-magnet is excited, greater attraction is set up and the permanent magnet is overcome. After the armature has thus advanced one-quarter of a turn the circuit is broken, and the next quarter of a turn is taken up and completed by the permanent magnet, which again comes into evidence. So that at every contact made by the primary the snail-armature actually advances half a turn, the intervals shown on the dial depending upon the gearing of the centre arbor.

"Princeps" Secondary. A simple form of dial mechanism is the "Princeps," shown diagrammatically in Fig. 55. A wheel, W_1, of 60 teeth is mounted on the centre arbor of an ordinary motion work and gearing with it is an idle wheel, W_2. Projecting from the armature, *A*, is a pallet, *P*, which is normally held by means of the spring, *S*, in a space between two teeth of the wheel, W_1. When the electro-magnet, *M*, is energised and the armature attracted, the pallet crosses over to the other wheel and advances it half a tooth. In returning, after the circuit is broken, under the influence of the spring, *S*, the pallet, *P*, enters and advances the succeeding tooth of the wheel, W_1, which has already moved forward with the wheel W_2. In this

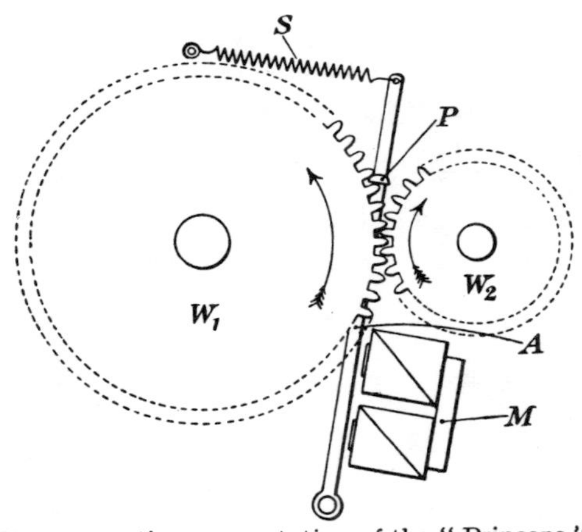

Fig. 55.—Diagrammatic representation of the "Princeps" secondary.

way, one complete tooth of the wheel W_1 is advanced with every signal from the primary.

In connection with multiple secondary working it may be of interest to refer here to a relay device introduced into the "Princeps" system. As explained on p. 115, an important feature of the "Princeps" primary is the polarised electro-magnet switch mechanism known as the "reverser." This reverser is again used in conjunction with a relay, the object of the latter being to keep the current supply to the primary distinct from that required for the secondaries.

In this case, there is a dial mechanism, similar to that described above, which is wired direct to the primary, but mounted on the centre arbor of this secondary is a snail, *S*, shown in Fig. 56. Two spring contacts, C_1, C_2, are normally held open with their outer springs, P_1, P_2, resting equidistantly apart on the edge of the snail. The contacts are connected to the respective solenoids of the reverser, *R*, in series with which is included the relay, *L*. With every half revolution of the snail, therefore, contact is made at C_1 and C_2, the current flows through the relay and the rocker

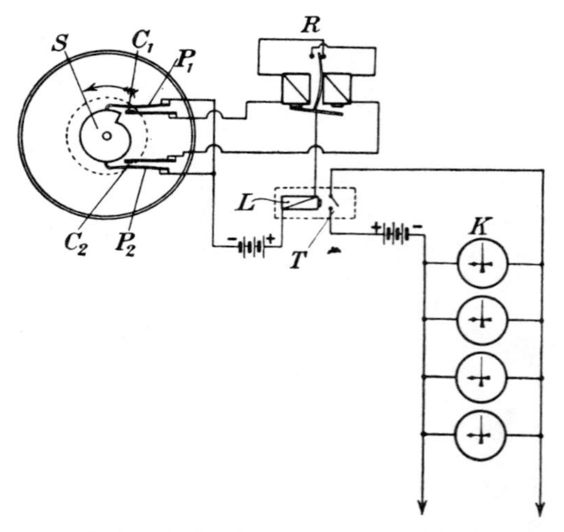

FIG. 56.—Relay device for working multiple secondaries ("Princeps" system).

armature of the reverser changes over, breaking the circuit in the process. The remaining secondaries, *K*, of the system are simply connected in parallel with the relay contacts, *T*, current being supplied from an independent battery.

Synchronising. Electrical attachments for the purpose of synchronising the hands of mechanical clocks were devised far back in the commercial history of electrical clocks. They have, however, been to a large extent superseded by the direct connection now formed between primaries and secondaries of the types just described. There are, nevertheless, two synchronising systems which demand attention, one originated by Dr. Hipp and the other by Messrs. Barraud and Lund. The latter is a simple device consisting of an electro-magnet which is excited hourly or at certain hours by a primary. Connected with the armature, which is attracted on these occasions, are two pivoted bell-crank levers from the extremities of which projects a pin, each coming through the dial on either side of zero. As the minute hand of the mechanical clock approaches the hour it enters the space between the pins. Directly the signal is received from the primary, the bell-crank levers close like a pair of scissors and, gripping the minute hand, reset it to the hour exactly. Any slight inaccuracy, either fast or slow, is thus immediately corrected.

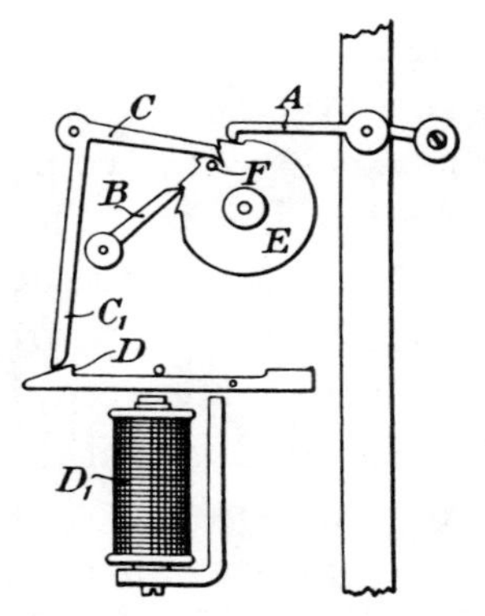

FIG. 57.—" Pulsynetic " waiting-train system for synchronising turret clocks.

Dr. Hipp's system is very similar, except that an inverted V-forked gravity lever, which descends sharply on a pin at the appropriate moment, is used instead of the bell-crank levers acting direct on the minute hand.

What must be regarded, perhaps, as a more important development of synchronising is the system as applied to turret clocks, instituted for the purpose of overcoming troubles due to the exposure of large hands to inclement weather.

Fig. 57 illustrates the " Pulsynetic " system of synchronising for this purpose. The movement, which is described as a " waiting-train," is in reality a pendulum maintained on the " Hipp-trailer " principle (p. 121) which is rated to drive the motion work slightly in advance of the pre-determined signals of the primary. Assuming that these occur every half minute, the " waiting-train " covers the distance in about 27 seconds. The pendulum, in vibrating, pulls over a count wheel, *E*, carrying the pin, *F*, which at the end of each revolution of the wheel raises the bell-crank lever, *C*, C_1, over the notched armature, *D*, and into the path of the pawl, *A*, so that the wheel is subsequently prevented from advancing. At the precise instant of the half minute, the notched armature, *D*, is attracted at the signal from the primary to the coil, D_1, thereby releasing the lever, *C*, C_1, and liberating the pawl, *A*, to recommence advancing the count wheel through which the motion work is driven. The click, *B*, prevents back-action.

Many are the devices which may form part of a secondary circuit, but they are too numerous to be detailed here. Factory time recorders, idle-machine recorders, warning signals of different descriptions all come under this category, and the ever increasing present day demand for such instruments enhances enormously the importance of what, at the commencement of this chapter, it was desired to emphasise—namely, that electrical clocks were now to be regarded as having assumed a most prominent and prospective branch of the science of horology.

PART III—WATCHES

CHAPTER XI

TYPES OF MOVEMENTS

Introductory. The earliest portable devices for indicating the hour of the day were merely pocket sundials, commonly known as "dials." The first piece of mechanical apparatus for this purpose dates back to about the year 1500, when Peter Henlein, of Nürnberg, invented the ingenious little instrument, later developments of which became known as "Nürnberg Eggs." It may be of interest to insert the following passage * from a printed school book, written at the time of this innovation by Johann Cocleus (1479–1522), a native of South Germany and at one time head-master of a school at Nürnberg. This book was published at Nürnberg in 1512 (probably by Weissenburger) and forms an abbreviated edition, presumably for the instruction of scholars, of Pomponius Mela's *Cosmography*.

manie/Germanica quidem lingua: In qua cernere licet/urbium
distātias fluuiorūq; cursus/exactius certe/quā vel in Ptolomei
Tabulis. ¶ Inueniūt indies subtiliora. Etenī Petr⁹ Hele/Iuue-
nis adhuc admodū/opa efficit/q̄ vel doctissimi admirant Ma-
thematici. Nam ex ferro paruo fabricat horologia plurimis di-
gesta rotulis/q̄ quocūq; vertant/absq; vllo pōdere & mōstrant &
pulsant xl. horas etiā si in sinu Marsupio ve cōtineant. ¶ Et si
Musicū q; audires certamē i die S. Katherine/nō possis,pfecto
nō mirari/tot cantores tāq; colonas voces simul reperiri. Tres

Petrus Hele.

Certamen musicum.

* Reproduced by courtesy of the Trustees of the British Museum.

A free translation of this would read as follows :—

> "Every day more subtle devices are invented; indeed a young man, Peter Henlein, has constructed works which amaze the most skilled mathematicians; for from pieces of iron he makes horologies containing many wheels: these horologies may be carried in any position, having no weights, and even in the pocket of the jerkin or in the pouch they go for forty hours and strike."

The watches referred to here were "clock-watches," so called because they were arranged to strike the hours successively just as a clock. In shape they were cylindrical, about two inches in diameter, and slightly later specimens, dated about 1530, may be seen at the British Museum.

For many years the verge escapement—a small replica of the clock verge escapement—held sway, and in spite of the low degree of accuracy that can have been secured from these watches there was generally no expense spared in lavish embellishment upon both the movements and cases. Even the plainest were provided with a richly-pierced and engraved balance cock. There are innumerable relics in museums and private collections, where the cases and dials are either pierced, embossed, engraved, chased or even beautifully enamelled to depict some biblical or allegorical subject. This tendency to artistic workmanship seems to have permeated the whole continent of Europe, with the result that watches were a most expensive ornament only to be indulged in by the wealthiest classes.

The craft, however, underwent a change on the advent of the cylinder escapement at the beginning of the eighteenth century, and was almost completely revolutionised about a century later, when the advantages of the lever escapement, originated by Thomas Mudge, became fully realised. To-day, the lever escapement in its developed form stands predominant in watchwork above all other types.

Another notable factor in the evolution of the watch was the introduction, some fifty years ago, of the keyless winding mechanism. This obviated the unpleasant necessity to the wearer of carrying a loose key and of the trouble in fitting it properly on the winding square whenever the watch needed winding. There is, however, still a certain demand for Swiss key-winding watches from various sources.

Designation of Movements. Watches fall into quite a large number of categories for the purpose of distinguishing the different designs and sizes. By the cases they are known in the familiar terms Open-Face, Hunter or Half-hunter, but the movements are gauged according to the diameter of the bottom (or pillar) plate. English and American watches follow the notation originally adopted by the Lancashire movement makers, describing them as such and such a "size," 10-size, 14-size, 16-size, etc. The 10-size movement measures exactly $1\frac{1}{2}$ inches and the difference per size in either direction is $\frac{1}{30}$ inch (or 0·0333). Usually, the odd numbers are omitted and, nowadays, many of the intermediate even numbers too. Small American wrist sizes commence at 10 below 0, and proceed upwards, including 6×0, 3×0, and 0, then come the now less numerous lady's pocket sizes, whilst at the top of the scale 14-, 16-, 18- and 20-size complete the usual range of gentlemen's pocket watches. The Swiss, on the other hand, have always employed the direct gauging by lignes. The "ligne"* is one-twelfth of a French inch, and the "douzième," which in the past particularly was most extensively used by watchmakers for measuring the thickness of plates and other small dimensions, is one-twelfth

* The word "ligne" is often abbreviated by a treble dash, thus, 13‴.

of a ligne. The French inch is longer than our own, measuring 1·067 English inches, and Table II draws a contrast between the two notations for describing the extreme diameters of watch movements.

TABLE II.—INCH-EQUIVALENTS OF STANDARD SIZES OF WATCHES

English and American sizes	Swiss lignes	Equivalent English inches	
—	5	0·443	Extra small wrist sizes.
—	6	0·531	
—	7	0·620	
—	8	0·709	
—	9	0·797	
10 × 0	—	0·866	Normal wrist sizes.
—	10	0·886	
—	11	0·975	
6 × 0	—	1·000	
3 × 0	—	1·100	
—	13	1·152	
0	—	1·166	
2	—	1·233	Ladies' pocket sizes.
4	—	1·300	
—	15	1·329	
6	—	1·366	
—	16	1·417	
8	—	1·433	
10	—	1·500	
—	17	1·506	
12	—	1·566	
—	18	1·594	Gentlemen's pocket sizes.
14	—	1·633	
—	19	1·683	
16	—	1·700	
18	—	1·766	
—	20	1·772	
20	—	1·833	

The tremendous Swiss output of the small wrist sizes, in recent years, has led to considerable competition in actual dimensions and design. From 5''' up to 11''' particularly there is subdivision into $\frac{1}{2}'''$ and $\frac{1}{4}'''$, and in addition to small round movements, oval and rectangular shapes have become quite general.

With the exception of perhaps a few isolated pieces the English productions have never descended to the small calibres which have become such a feature of Swiss work. When ladies' pocket watches were in vogue, 2-, 4-, 6-, 8- and 10-size covered the usual range and anything smaller than 0-size is rarely met with. The American factories, however, have produced wrist-size models, though they do not specialise in these to anything like the extent of the Swiss.

CHAPTER XII

WATCH TRAINS

Calculation of Watch Trains.—In watches the time-keeping medium is the balance and balance-spring, in contrast to the pendulum of a clock. The balance performs the arcs or vibrations of some definite duration just as the pendulum, but much more frequent. The vast majority of watches in general use have balances which make 18,000 vibrations, to and fro, in an hour—that amounts to 5 every second. Each vibration releases a tooth of the escape wheel from one of the two pallets, so that it will easily be seen that an additional wheel and pinion is necessary in a watch to the number found in a clock, to meet the needs of the more frequent vibrations. This is known as the "fourth wheel and pinion" and, in most cases, it is designed to make one revolution in a minute. The pinion is often provided with a long pivot to come through the dial and carry a seconds hand. Taking the case of the 18,000 train, the escape wheel has 15 teeth, and as each tooth is checked in turn by both pallets it takes 30 vibrations of the balance to allow the wheel to make a complete rotation, which is equal to one-tenth of a minute or 6 seconds. The fourth wheel, if required to go round once a minute, has, therefore, to give a ratio of 1 to 10—which means that, given an escape pinion of 6, 7 or 8, a fourth wheel is required of 60, 70 or 80 respectively. In the case of a 14,400 train, which beats 4 to the second instead of 5, the ratio is 1 to 8, so that, with an escape pinion of 6, 7 or 8, the fourth wheel is 48, 56 or 64 respectively. In the

16,200 train the balance makes 4½ vibrations per second and cannot, therefore, show true seconds, though the fourth wheel can make one rotation in a minute. The ratio is 1 to 9 and, with an escape pinion of 6, 7 or 8, the fourth wheel is 54, 63 or 72 respectively.

The fourth wheel in most cases conforms to one or other of the above conditions and thus makes 60 revolutions to 1 of the centre, but there are exceptions and it is better to consider the equation covering the whole system, as in dealing with clock trains (see p. 58), thus :—

$$\left.\begin{array}{l}\text{Vibrations of}\\ \text{balance per hour}\end{array}\right\} = \frac{2\,(\text{Centre} \times \text{3rd} \times \text{4th} \times \text{escape wheels})}{\text{3rd} \times \text{4th} \times \text{escape pinions}}$$

abbreviated to :—

$$\text{Vibs.} = \frac{2CTFE}{tfe}.$$

This formula becomes of very practical assistance in the case of any wheels or pinions having been lost.

Table III shows a few of the combinations most frequently met with.

TABLE III.—COMBINATIONS FOR WATCH TRAINS

Centre wheel	Third pinion	Third wheel	Fourth pinion	Fourth wheel	Escape pinion	Beats per hour	
64	8	60	8	60	6	18,000	Usual in Swiss horizontals
64	8	60	8	70	7	18,000	General seconds trains
64	8	60	8	80	8	18,000	
80	10	75	10	80	8	18,000	
80	10	75	10	72	8	16,200	High numbers
64	8	60	8	63	7	16,200	Usual numbers
64	8	60	8	60	7	15,428	Old style Eng. levers
64	8	60	8	56	7	14,400	

Motion Works.—In watches the calculation for motion works is the same as in clocks, excepting that the cannon pinion and minute wheels are never identical in size and number.

Table IV, whilst not exhausting all the possibilities, includes many combinations of motion works which are in general use.

TABLE IV.—COMBINATIONS FOR MOTION WORKS OF WATCHES

Hour wheel	Minute nut	Minute wheel	Cannon	
42	14	48	12	"Waltham" 16- and 18-size.
40	10	36	12	
48	8	32	16	
32	8	24	8	
32	8	27	9	
27	6	24	9	Unusual 5½ lignes Swiss.
32	8	30	10	"Waltham" o-size.
40	8	24	10	
48	10	30	12	
36	8	32	12	
48	8	28	14	
42	7	30	15	

Movement Designing and Planning.—Figs. 58–66 give a comprehensive idea of the evolution in planning of watch movements. The variety in Swiss designs to-day is legion, and to a great extent follows the public demand for variety combined with utility and cheapness. The greatest care is usually taken in watch factories in scheming out the details of a new model and, as soon as it has been made up and its worth proved, mass reproduction on a large scale follows. This system, however, does not altogether apply to the higher grades, which must undergo a certain amount of hand finishing and adjusting. In England mass production of watches has, perhaps unfortunately, never been developed on similar lines to that prevailing in Switzerland.

It is to be feared that the prosperity of fifty years ago in English watch manufacturing by hand methods promoted a most inflated spirit of self-satisfaction, which induced these successful pioneers to turn an obstinate ear to any proposal involving the adoption of repetition machinery. They were powerful enough to drive even would-be adventurers in such a new departure out of the country, a policy which undoubtedly resulted in the birth of the American factories.

One has only to turn to the first volume of the *Horological Journal*, published in 1858, in order to substantiate this fact. Following an interesting account of the inauguration of the factory at Waltham, Massachusetts, a correspondent writing to the Editor says :—

> "In any other publication than the *Horological Journal* I should have supposed the account of the Factory at Waltham, Massachusetts, to have been one of those alarming statements occasionally put forth for trade purposes of the Puff School, intended to create a panic, to be taken advantage of in some way or other, as I doubt not this account was originally. . . . There can be no doubt that the production of the machine may be simplified to a great extent by the adoption of a limited number of sizes and the employment of the going-barrel form; but this, with the lever form of movement, will leave the machine, however pretty, with but an indifferent character for that which should be the *sine quâ non* in all, viz., the faculty of keeping time.
>
> "Let, therefore, Brother Jonathan go ahead, and create by any means that please him a continually increasing demand for good watches, and I feel confident that it will still be England from which a large part of the corresponding supply will flow."

A name which will ever be remembered in association with the introduction of machines for reproducing watch parts is that of Pierre Frederick Ingold (1787–1878). A Swiss by birth, Ingold worked for some time with Bréguet in Paris, but afterwards came to London and assisted in starting the British Watch Company. This concern failed within about two years of its inception through the opposition it received from the orthodox industry of that period, and the nucleus passed to America. Ingold himself went to New York in 1845, only five years after coming to London.

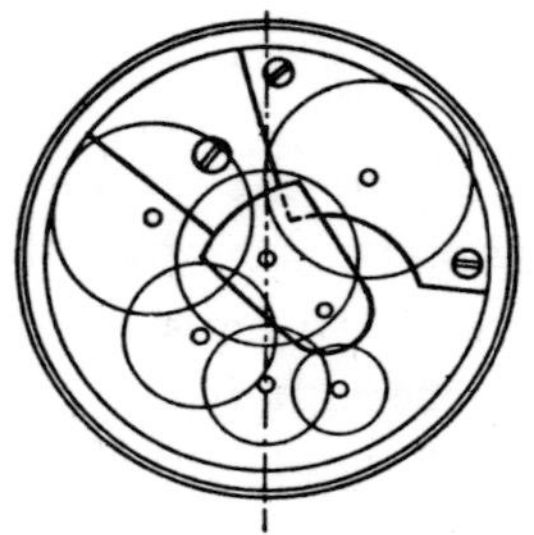

FIG. 58.—Diagram showing the calibre of an early English full-plate keywind movement.

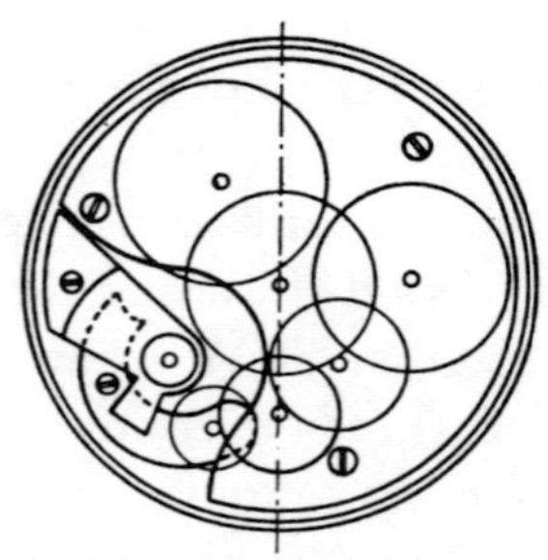

FIG. 59.—Diagram showing the calibre of an English ¾-plate keywind movement.

The fact must not be overlooked, however, that when the Americans and Swiss had begun to show the lead, valiant efforts were made to produce machine-made watches in England at comparatively low prices. Possibly because the commencement was made too late and technical details had not received sufficient consideration, or perhaps because prices were cut too low in the hope of creating a market, most of these productions very regrettably never gained any ascendancy and some have already become extinct.

The Swiss, fifty years ago, cannot be said to have been any more advanced in the art of watch manufacturing than the English, but they moved with the times, studied closely the technical side, evolved intricate machinery—in short did everything possible to produce a watch that would go properly at considerably less cost than a hand-made watch. They have succeeded and have achieved world-wide supremacy in the industry.

The first model to be adopted for general use in this country after the introduction of the lever escapement was the full-plate, more or less following the style of the old verges, as shown in Fig. 58. Later, the ¾-plate shown in Fig. 59 appeared as an improvement, reducing the thickness of the watch. When the keyless mechanism was perfected

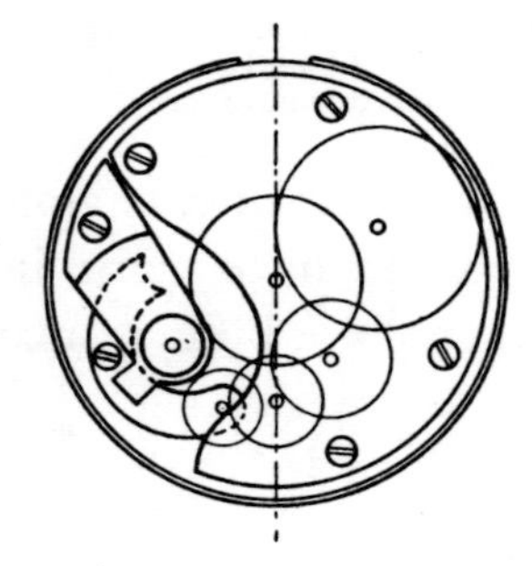

FIG. 60.—Diagram showing the calibre of an English ¾-plate keyless open-face movement.

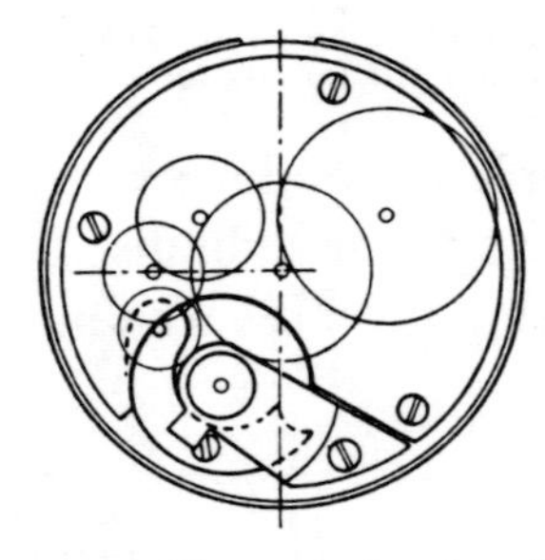

FIG. 61.—Diagram showing the calibre of an English ¾-plate keyless hunter movement.

and the going barrel had superseded the fusee and chain, the approved design became as that of Fig. 60. This arrangement of the train is adapted to the requirements of the open-face case, but Fig. 61 indicates the same movement for use with a hunter case.

Although numerous individual pieces of different designs are met with, there has been no variation in this general form of English movement since its introduction to the present time.

Fig. 62 shows an early example of Swiss machine-made

work, namely, the keywind horizontal (or cylinder) bar movement, and Fig. 63 marks the advent of keyless mechanism to a movement of the same kind. Later, it became

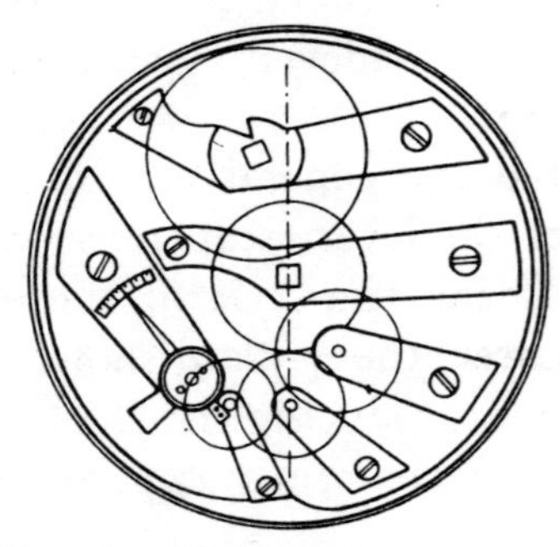

Fig. 62.—Diagram showing the calibre of an early Swiss keywind cylinder bar movement.

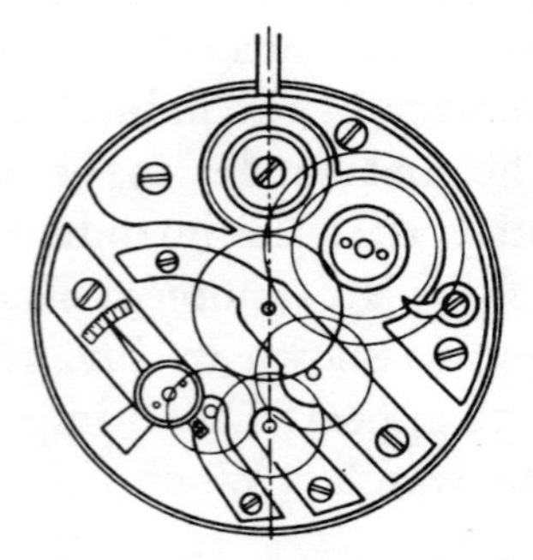

Fig. 63.—Diagram showing the calibre of a Swiss keyless cylinder bar movement.

possible to produce watches with lever escapements almost as cheaply as cylinders. Gradually these have come into amazing prominence and the movements are built in many

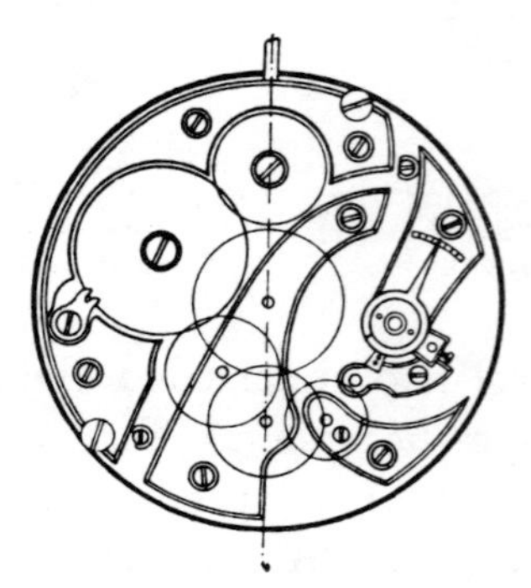

Fig. 64.—Diagram showing a typical example of the calibre of a modern Swiss lever movement, adaptable both to pocket and wrist sizes.

different ways, of which Fig. 64 may be regarded as typical of both large and small sizes. To-day the cylinder watch is no longer popular except in cheapest grades, the pocket watch for ladies has practically vanished and the demand for small wrist movements has grown enormously. Figs. 65 and 66 may be regarded as representative of this class,

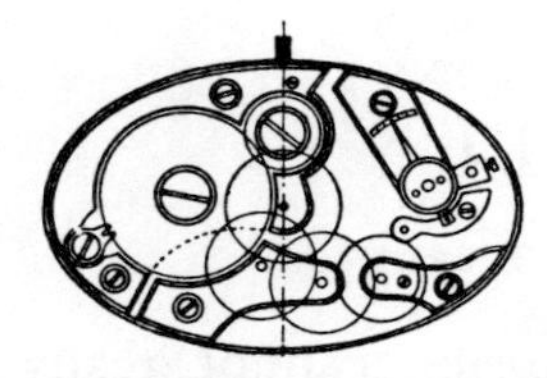

FIG. 65.—Diagram showing the calibre of an oval lever wrist movement.

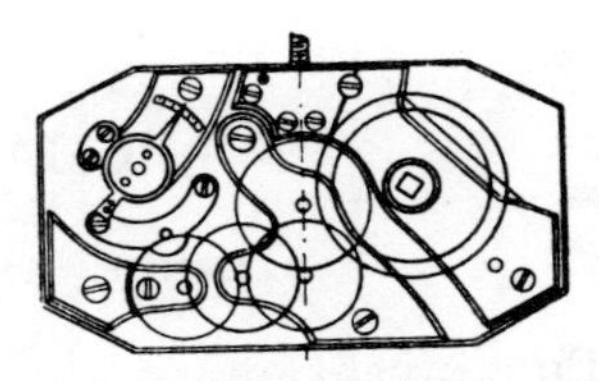

FIG. 66.—Diagram showing the calibre of a rectangular wrist movement.

though here again the actual varieties in design adopted by different makers are extremely numerous.

CHAPTER XIII

BALANCES AND BALANCE-SPRINGS

THE invention of the balance-spring cannot really be ascribed to any one man, though it is probable that it was largely the outcome of theories demonstrated by the scientist, Dr. Hooke, towards the end of the seventeenth century.

The earliest watch balances were "foliots," similar to those used in clocks and described in Chapter II. The weights, however, were fixed at the extremities of the arm and generally were spherical in form. The motion of the foliot was controlled by a hog's bristle fixed at one end to the top plate, across which it extended diametrically. At the free end a short projection at right angles entered the path of the foliot. When struck by the edge of a weight the springiness of the bristle, which could be increased or decreased by means of a slide, provided impetus for the succeeding vibration.

Next to foliots came balances, which were very thin, three-armed, steel wheels. Later, the same form appeared in brass and, still later, the practice of rounding the rim on top was introduced. This design, known as the plain balance, was, in the best watches, made of gold. The "compensation balance" is a more recent development devised to neutralise the effect of temperature upon both the spring and the balance.

From the time-keeping standpoint, the effect of varying temperature upon the balance-spring is the principal cause of error. Such fluctuations play havoc with its elastic property. When the temperature rises, the balance-spring becomes less energetic and cannot do its work so briskly, but, conversely, in a low temperature its performance is more vigorous. This fact should be carefully noted, as it is a far greater cause of error in time-keeping than any other factor. Error due to the increase or decrease in the dimensions of both balance and balance-spring through change of temperature has only a small effect on time-keeping.

The time of vibration of a balance is governed by the moment of inertia of the balance and by the bending moment of the balance-spring, and these two features must be carefully investigated both individually and collectively in order to understand, first, the actual effect varying temperature has upon the time of vibration and, secondly, the remedy of compensation.

The motion of a balance is "harmonic," even more truly so than that of a pendulum, but whereas, in the case of the pendulum, gravity is an important component factor, it plays no part whatever in controlling the vibration of the balance.

The property in matter which opposes commencement or variation of motion is called "inertia" and it is proportional to the mass of the body. It is embodied in Newton's First Law of Motion, which says that "Every body remains in its state of rest or of uniform motion in a straight line except in so far as it may be compelled by forces to change that state."

The freedom with which a balance vibrates depends considerably upon its inertia. A very heavy one would offer greater resistance to motion and require greater force to start it in motion than a lighter one and, once started, there would be a correspondingly big force needed to stop it, assuming that the acceleration is the same in both cases.

A balance rotates about an axis perpendicular to its plane, consequently its rate of motion is angular velocity, and its rate of change in velocity is angular acceleration.

In the case of a body moving along a straight line, the force impelling the body, or the resistance the body offers to motion, equals the mass of the body multiplied by the acceleration, but in a rotating body the moment of the force equals the moment of inertia, I, multiplied by the angular acceleration, α. Supposing the body to consist of many particles of masses, m_1, m_2, m_3, etc., at distances r_1, r_2, r_3, etc., from the axis of rotation, then the resistances offered by the particles will be $m_1r_1\alpha$, $m_2r_2\alpha$, $m_3r_3\alpha$, etc., α being the angular acceleration, and the moments of resistance or the "turning moments" will be $(m_1r_1\alpha \times r_1)$ or $m_1r_1^2\alpha$, $m_2r_2^2\alpha$, $m_3r_3^2\alpha$, etc.

It will be seen that the angular acceleration is a common factor, so that the total resistance offered by the body will be $\alpha(m_1r_1^2 + m_2r_2^2 + \ldots\ldots)$. The expression in brackets is the moment of inertia, I, of the body, comprising the sum of the moments of inertia of each separate particle.

Thus the moment of inertia varies as to the square of the distance of the particle from the axis of rotation. Suppose that the total mass, M, of the rotating body be concentrated at a radius k such that $Mk^2 = I$, then the radius k is called the "radius of gyration" and, by transposing the above expression :—

$$k = \sqrt{\frac{I}{M}}.$$

Such are the characteristic properties of the balance ; now let us turn to the balance-spring. It must not be forgotten that the behaviour of the elastic property of the spring under varying temperature is the principal cause of inaccuracies in the time of vibration of the balance.

Elasticity enables a material to undergo distortion

without yielding to a permanent change of form on the removal of the forces producing distortion. If a material when depressed or loaded by stress reaches a stage when it refuses to return to its original form, it is said to have been stressed beyond its "limit of elasticity" and to have received a "permanent set."

Stress is the force per unit area of section which produces distortion in any material. As a formula it is expressed by :—

$$\text{Stress} = \frac{\text{Load } (W)}{\text{Area } (a)}.$$

Under the influence of stress a rod or wire may be strained or stretched longitudinally and the term strain is used to describe the ratio between the elongation and the original length of the material, *i.e.* :—

$$\text{Strain} = \frac{\text{Elongation } (e)}{\text{Initial length } (L)}.$$

Consideration must now be given to what is perhaps the most important factor. Dr. Hooke advanced the theory, which has now become known as "Hooke's Law," that : "Strain is proportional to the stress producing it." Assuming that we are concerned with a material of uniform cross-section, such as a rod, which is acted upon by stress so that there is longitudinal strain, then the ratio of the stress to the strain produces a result called the Modulus of Elasticity (E). This is commonly known as "Young's Modulus" because it was first enunciated by the scientist, Thomas Young, and it is thus also distinguished from other moduli which do not enter into the present subject. The modulus of elasticity for any given material at a given temperature may be expressed thus :—

$$E = \frac{\text{Stress}}{\text{Strain}} = \frac{W/a}{e/L} = \frac{WL}{ae}.$$

The value of Young's modulus of elasticity varies according to the resistance any particular material offers to stretching; thus the value for steel is about 13,000 tons per sq. in., whilst that for wood, such as oak or fir, is about 700 tons per sq. in.

Returning to Hooke's Law, it can be shown that if an elastic material, acting as a spring and being rigidly fixed at one end, is wound through any given angle, then the bending moment is directly proportional to the angle of winding.

In a balance-spring, one end is held at the stud whilst the other follows the movement of the balance and the spring is alternately wound and unwound.

The formula for determining the bending moment of a material of rectangular section must here be accepted without explanation, as that is not possible within the scope of this work.* If, however, A represents the angle through which such a spring of rectangular section is wound, then the :—

$$\text{Bending moment } (\beta) = \frac{Ebt^3A}{12L},$$

L, b and t being the length, breadth and thickness respectively and 12 the constant applied to materials of rectangular section. A is measured in radians.

It is now possible to proceed to examine the theory of the time of vibration of a balance, combining the components of both balance and spring.

It has been mentioned that the motion of the balance is harmonic, though more perfectly so than that of the pendulum.

* Readers desiring further information on this and associated matters may be referred to *The Strength of Materials* by Ewart S. Andrews (Chapman and Hall), 13*s*. 6*d*. net.

The harmonic formula for a single vibration is :—

$$\mathcal{T} = \pi\sqrt{\frac{\text{angular displacement } (A)}{\text{angular acceleration } (\alpha)}}.$$

Angular acceleration, which has been referred to before in consideration of the moment of inertia of a rotating body, is equal to the turning moment divided by the moment of inertia. Thus :—

$$\alpha = \frac{\beta}{I}.$$

As A is the angle through which both the spring and balance are wound, it must also represent the angular displacement, so the expression becomes :—

$$\mathcal{T} = \pi\sqrt{\frac{A}{\beta/I}}$$

$$= \pi\sqrt{\frac{AI}{\beta}}.$$

It follows, however, that the turning moment of the balance must be identical with the bending moment of the spring. Hence :—

$$\beta = \frac{Ebt^3A}{12L},$$

$$\therefore \qquad \mathcal{T} = \pi\sqrt{\frac{IA12L}{Ebt^3A}}$$

$$= \pi\sqrt{\frac{I12L}{Ebt^3}}.$$

The angle of winding cancels out and disappears as a factor governing the time of vibration, but those left comprise π, the moment of inertia of the balance and the dimensions and modulus of elasticity of the spring.

Let us carry this formula further :—

$$\mathcal{T} = \pi\sqrt{\frac{I\,12L}{Ebt^3}},$$

but
$$I = mk^2,$$

$$\therefore \qquad \mathcal{T} = \pi\sqrt{\frac{mk^2\,12L}{Ebt^3}}.$$

In this expression, the terms E, t and k are sensitive to changes of temperature, whilst the other factors π, m, 12 and $\frac{L}{b}$ remain constant,

$$\therefore \qquad \mathcal{T} \propto \sqrt{\frac{k^2}{Et^3}}.$$

In order to maintain a constant time of vibration under any condition of temperature, thus eliminating the principal causes of error, it is the behaviour of the three terms k, E and t that has to be considered in endeavouring to arrive at some adequate means of compensation.

From a practical standpoint, it is possible to obtain extraordinarily good results, but it must be admitted that the final process particularly is rather through the expedient of trial and error than through construction based on mathematical calculations. The radius of gyration is a hopelessly awkward dimension to arrive at in practice, just as the centre of oscillation is in a pendulum bob, but it can be determined approximately and care is taken that the balance shall consist as far as possible of rim only, so as to concentrate the mass of the balance therein.

The modulus of elasticity of the spring, changes in which may be regarded as the chief cause of error, is equally difficult to determine under varying conditions of temperature, especially as its fluctuation is not uniform and its value may not be precisely the same for different specimens of the same material.

These matters, however, will be referred to again in dealing with compensation balances. Of the remaining factors, the length of spring and the mass of the balance should be noticed. From the formula, the time of vibration is proportional to $\sqrt{m}$ and $\sqrt{L}$, thus

$$T^2 \propto m \text{ and } L.$$

Mass is proportional to weight, so that

$$T^2 \propto \text{weight}.$$

The effect of this is that a balance can be brought to time, neglecting temperature, by adjusting the length of the spring or altering the weight of the balance.

The following examples may serve as a practical illustration :

Example 1. A watch is losing half an hour a day with a balance weighing 1·2 grammes. What should the correct weight be ?

The balance is too heavy and only 23½ hours are recorded instead of 24, thus :—

$$w : 1{\cdot}2 :: (23{\cdot}5)^2 : (24)^2;$$

$$w = \frac{(23{\cdot}5)^2 \times 1{\cdot}2}{(24)^2}$$

$$= \frac{662{\cdot}7}{576}$$

$$= \underline{1{\cdot}15 \text{ (approx.) grammes.}}$$

Example 2. A watch is gaining half an hour a day with a balance-spring 240 mm. long. What should the correct length of spring be ?

The spring is too short and must be " let out." The watch is recording 24·5 hours instead of 24, thus :—

$$L : 240 :: (24{\cdot}5)^2 : (24)^2;$$

$$L = \frac{(24{\cdot}5)^2 \times 10}{24}$$

$$= \frac{6002{\cdot}5}{24}$$

$$= \underline{250 \text{ mm. (approx.).}}$$

It will be seen, therefore, that a plain balance consisting only of a rim and very slender arms and centre, combined with a balance-spring of hardened and tempered steel or other suitable material, is seriously exposed to these disturbing influences, modulus of elasticity and radius of gyration, which so considerably affect the time of vibration

In the early days of horology it was thought, and indeed it is still a question often raised, that if the vibrations squared are proportional to the length of the spring, expansion or contraction of the spring should cause conspicuous error in the rate. This theory was, however,

FIG. 67.—Early form of plain balance.

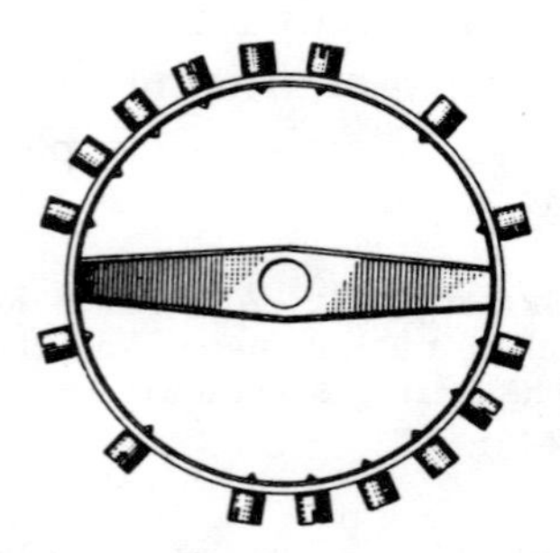

FIG. 68.—One form of Swiss plain balance. Although a bimetallic rim is used, the balance is uncut and does not provide a means of compensation.

disproved in 1882 by Mr. T. D. Wright, who has been for many years the highly-esteemed Head of the Joint Horological Department of the British Horological and Northampton Polytechnic Institutes. Mr. Wright pointed out that temperature variations affecting the length of the spring affect also the breadth, but with directly opposite effect, so that the net result is zero; the ratio $\frac{L}{b}$ can be regarded as a constant and not contributory to errors in time-keeping.

As regards the thickness; this increases under expansion

and aids compensation by producing a contrary effect to the change of modulus. Whilst a decreasing modulus, following a rising temperature, causes the spring to become weaker, the simultaneously increasing thickness actually tends to strengthen it.

Fig. 67 shows the form of plain balance in vogue for many years, even in expensively constructed watches. To-day, however, a large percentage of watches are fitted with what appears to be a compensation balance as shown in Fig. 68, namely, a bimetallic rim provided with screws, but uncut or partially cut, which behaves in precisely the same way as a plain balance. The only virtue which may be claimed for this form, apart from cheapness, is that under certain circumstances the watch may be brought to time by altering the weight of the balance by means of the screws and timing washers.

Compensation Balances. It has been shown that, through change of temperature, varying modulus is the chief cause of error in time-keeping. The order of the error in the time of a steel spring and plain balance under the influence of rising temperature is a loss of about 11 seconds per day per degree Centigrade. With an increase of temperature the modulus decreases and the spring weakens, the time of vibration increases and the watch loses, so that, to counteract this, the balance must contract or decrease in size. The compensation balance is designed to perform this function automatically.

Fig. 69 shows a customary form of such a balance for watches. It consists of a bimetallic rim, the inner lamina and crossbar being steel and the outer lamina brass. The rim is cut through obliquely and at positions diametrically opposite, near the crossbar. At uniform intervals round the circumference it is drilled and tapped to take screws with large heads which act as weights and can be distributed in the rim as adjustment for compensation demands. The two metals, having different coefficients of expansion, that of the brass on the outside being greater, produce, under the influence of increasing temperature, the effect of curving the rim inwards at the free end, the maximum amount being near the cut. Conversely, a decreasing temperature causes an outward movement of the rim. The oblique cuts in good balances protect the rim from losing its form whilst under

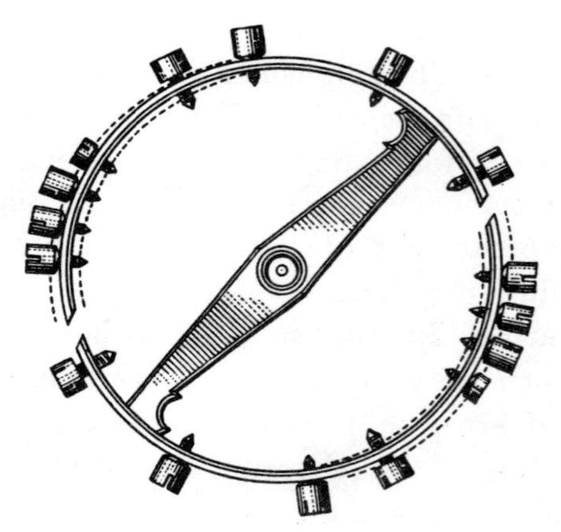

FIG. 69.—A bimetallic compensation balance.

manipulation. In watch balances there are generally sixteen or eighteen screws and about thirty holes, and the positions the screws are required to occupy is determined by putting the watch through a series of tests at different temperatures. The screws may be gold or brass. If a palladium spring is employed it is sometimes necessary to use platinum screws to give extra weight when all the screws are assembled near the cut and there are no vacant holes. Used under any other circumstances platinum screws are only an ornament.

In practice a watch is over-compensated if it is losing in

the ice-box and gaining in heat and the remedy is then to group the screws away from the cut towards the fixed end. Conversely, if there is a loss in heat more compensation is needed and the grouping must be towards the cut. If the weight is concentrated near the fixed end less mass follows the movement of the rim under expansion and contraction, and less variation of the radius of gyration ensues.

It is the automatic change in the radius of gyration, therefore, which provides a means of correcting as far as possible the variations in the modulus of elasticity of the spring, as well as the very small error due to variation in the volume of the balance.

The appropriate ratio of brass and steel used in the bimetallic balance needs some consideration. Exhaustive experiments have been carried out by scientists to arrive at the most desirable combination responding to all conditions. These efforts have led to the usual practice of making the thicknesses of brass and steel in the proportion of 3 to 2, an important feature being that the materials shall offer equal resistances to bending, so that the change of curvature may be uniform. If this is not carried out there would be a tendency for the two metals to assume different lengths where their surfaces are united and then maximum sensitivity would not be attained.

This may be more apparent if viewed from another standpoint. The neutral axis of a rectangular bar is the longitudinal section midway between the two edges, which remains the original length after the bar has been bent into a curve. Under such conditions the outer edge of the curve is extended beyond the length when it was straight, whilst the inner edge is reduced by a similar amount and the strain at any point in the bar is proportional to its distance from the neutral axis. Assuming the bimetallic rim of a balance to consist of brass and steel of different thicknesses, then the neutral axes of the two metals are separated by a distance equal to half the total thickness of the rim. If under the influence of varying temperature the change of curvature can be such that the relative position of the respective neutral axes in taking up the new dimensions is not altered then the individual curvature of the two metals will correspond and the greatest sensitivity will result. With the brass outside the steel, the process of bending has an impelling effect upon the expansion of the outer edge of the steel where the two metals are united and a correspondingly restraining effect upon the expansion of the inner edge of the brass, so that these two edges will remain equal in length. This condition can only be brought about by the two metals offering equal resistances to bending. If there is a preponderant proportion of one metal over the other, as for instance brass much thicker than steel, the change of curvature would be extremely small because there is no inherent desire for the metals to expand or contract otherwise than in a straight line. The thick brass would override the weak inclination of the thin steel and the balance would not be at all sensitive.

Much of the scientific research on this subject was carried out by Yvon Villarceau who wrote a treatise on the subject, entitled: *Recherches sur le mouvement et la compensation des chronomètres* (1862). He claimed that the brass and steel should be in the ratio of 17 to 12 and that if the bending moments are to be equal, the :—

$$\frac{\text{Thickness of brass}}{\text{Thickness of steel}} = \frac{\sqrt{E} \text{ of steel}}{\sqrt{E} \text{ of brass}}.$$

The actual movement of the rim depends upon its expansion and contraction so that in effect a relationship is then being created between the modulus of elasticity of the spring and the coefficient of expansion of the balance.

Middle-Temperature Error. Despite the accuracy with which the foregoing bimetallic formula may be applied in practice there is still a fault which causes inconsistency in the rate. It is possible that the compensation may give very satisfactory results in the extremes of a range of temperature and poor results in the mean temperatures, hence the designation "middle-temperature error." The amount of this error can only be observed by experiment and, as it is more apparent when the mechanism has to be subjected to a wide range of temperature, it is perhaps of more material consequence in the case of "marine chronometers" than in watches, which can generally be confined within narrower limits of temperature.

The actual reason for this discrepancy, first noticed by Ferdinand Berthoud in 1775 with curb compensation (cf. Chapter XX, p. 244), and later by Ulrich, Dent and others with marine chronometers, has given rise to much speculation in the past and many efforts by scientific and practical men to devise some means of eradicating it.

The researches of recent years by the renowned scientist Dr. C. E. Guillaume, the inventor of Invar (see p. 29), have thrown much light on this problem. Dr. Guillaume points out in his work *La Compensation des Horloges et des Montres* that the fact of a metal weakening under increasing temperature is undoubtedly caused by the separation or dispersion of the molecules through expansion. Resistance to bending is, in fact, a result of intermolecular reaction and is least when the molecules are the most dispersed. The order of the dimension of elasticity is, nevertheless, quite different from that of expansion. Taking α to represent the coefficient of expansion and η the thermo-elastic coefficient, then for a given change in the modulus of elasticity, η will be in the order of 20 to 30α.

The progressive nature of these two factors, expansion and elasticity, when a material is submitted to an increase of temperature may be shown graphically as in Fig. 70A. Algebraically, it is possible in most cases to express the law of expansion by a quadratic equation, which can be written thus :—

$$L_\theta = L_0(1 + \alpha\theta + \beta\theta^2).$$

In this equation, the length L_θ of a bar at temperature θ

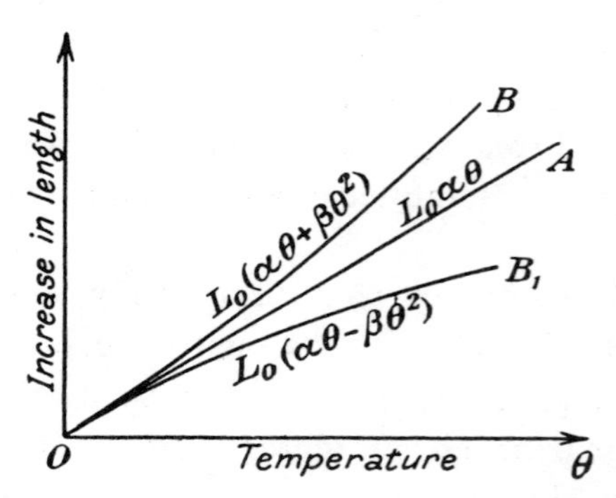

FIG. 70*a*.—Illustrating linear and quadratic laws of thermal expansion.

is obtained from the product of its length L_0 at 0° by the quantity contained in the brackets where α and β are respectively the coefficients of the linear term and the squared term. The tangential inclination of the curve at the point corresponding to any temperature θ_1 is $\alpha + 2\beta\theta$, which one regards as the true expansion at θ_1 and is equal to the mean expansion between 0° and $2\theta_1$.

If the linear term $(\alpha\theta)$ existed alone, the expansion of a bar would be represented by the line *OA* in Fig. 70A. According as the squared term $(\beta\theta^2)$ is positive or negative, the resultant expansion of the bar follows the curve *OB* or OB_1.

An equation of the same form expresses the variation in the modulus of elasticity with changing temperature, α and β being replaced by corresponding coefficients of elasticity.

In metals and alloys which behave normally the two coefficients of expansion are positive whilst those of elasticity are negative.

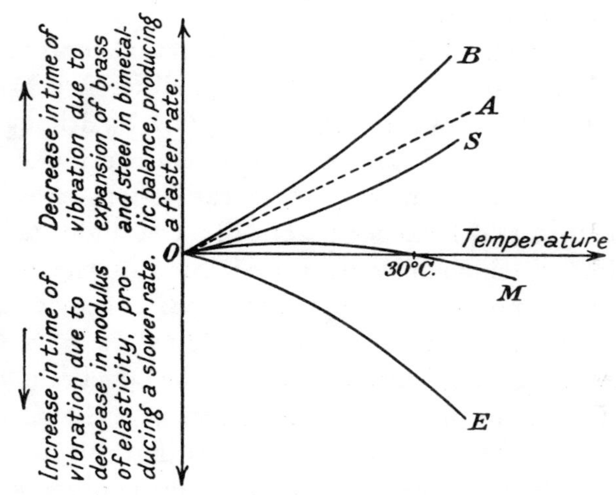

FIG. 70*b*.—Comparing the effects of the thermal expansion of brass and steel with the effects of thermal change in the modulus of elasticity of a steel balance-spring.

In order to obtain perfect compensation or, in other words, a constant rate at all temperatures, it is necessary to ensure that the mechanical means adopted for automatically counteracting the natural variation, is such that every detail appertaining to the cause of error is carefully taken into account. That is to say, it is not sufficient to settle on a means of compensation based on linear thermal expansion alone ; the quadratic law must also be followed. If this condition cannot be attained with uniformity compensation will be incomplete, and the rate will show the middle-temperature error common to most forms of compensation.

This fact is illustrated in Fig. 70*b*, where over a range of temperature between 0° and 30° C., the effect of a decreasing modulus of elasticity in a steel balance-spring is shown by the curve *OE*, whilst the effect of expansion on the brass and steel in the compensation is indicated by *OB* and *OS* respectively. The effect of expansion on these metals combined, which represents the actual compensation factor of the balance, is the algebraic difference between the two and is shown by the curve *OA*. Although the relative inclination of *OB* and *OS* differs considerably the squared terms in their respective expressions are almost equal, so that *OA* becomes practically a straight line, the inclination of which may be varied but not the form by altering the dimensions of the bimetallic rim or by changing the position of the rating screws. The rate of a watch thus compensated is determined by the algebraic mean, *OM*, between the factor of error, *OE*, and the factor of compensation *OA*, so that, by comparing this with the true mean, the uniformity of the rate can be observed. It will be noticed that *OM*, coincides with the true mean at *O*, the one extreme of the temperature, and cuts it at the other, shown as 30°, but at intermediate temperatures there is a distinct departure from the true mean. This fact indicates at once that the rate is affected through incomplete elimination of the squared value in the respective expressions and accounts for “ middle-temperature ” error which at its maximum amounts to about 2 secs. per day with this form of compensation.

Dr. Guillaume further demonstrated that the error was lessened considerably by employing a bimetallic rim of brass and nickel-steel in place of brass and steel. This discovery he made in 1899 and the results therefrom proved most encouraging.

As watches only are being dealt with at this stage, further reference to the subject of "auxiliary compensation" devised to overcome the middle-temperature error will be found in Chapter XX on Marine Chronometers.

Dr. Guillaume's researches have been continued, however, and in 1913 he invented an alloy of nickel-steel with a percentage of chromium, which he has named "Elinvar"

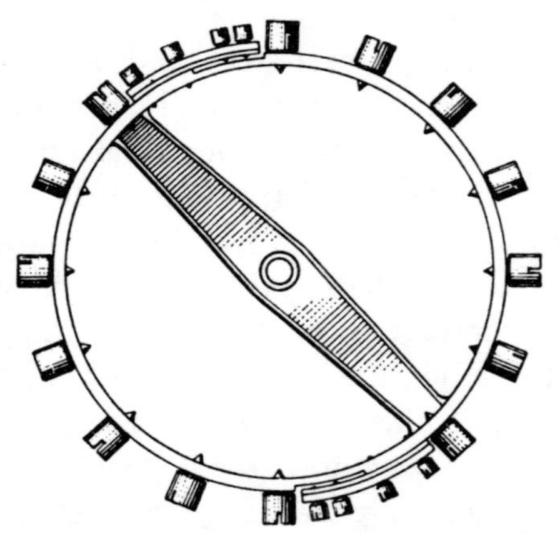

FIG. 71.—Paul Ditisheim's "Affix" monometallic balance with bimetallic compensation curbs for use in conjunction with an elinvar balance-spring.

(derived from "Elasticity invariable"). During recent years, the worth of this wonderful alloy has been and is still being proved as a material for balance-springs. It shows a practically negligible thermo-elastic coefficient, and not requiring, therefore, any compensation presents a very different problem with regard to the balance. Paul Ditisheim has overcome this difficulty by introducing a non-magnetic balance, which is uncut and the small amount of compensation required by the balance itself is met by adding to the rim, as shown in Fig. 71, two short bimetallic curbs.

The results obtained from watches provided with this balance and spring combination are exceptionally gratifying, and it seems likely that this recent contribution to the science of horology may revolutionise the practical side of the, hitherto, somewhat complex and elusive question of compensation for temperature.

The Balance-spring. Steel hardened and tempered blue is the material most frequently used for making balance-springs. Palladium alloy, however, is met with, especially in marine chronometers, as well as certain other white alloys, notably nickel-steel and nickel-iron (ferro-nickel), whilst " Elinvar " must now also occupy a prominent position in this class. Of this group some are magnetic and some non-magnetic, yet to all appearances they are identical. An enormous number of Swiss watches are now fitted with white springs other than palladium and, with normal changes of temperature, it is claimed that they give better results than steel when used in conjunction with an uncompensated balance. This may be due to their possessing a lower coefficient of expansion and a smaller variation in modulus. In practice, palladium being tough and nickel alloy springs being very soft, they are not so pleasant to work as steel. Palladium is a somewhat rare mineral and the principal alloy used for balance-springs was invented by C. A. Paillard about 1879. The only advantages it possesses are those of not rusting and being non-magnetic.

Isochronism, which has been previously shown as such an important factor governing the arcs described by a pendulum, once again appears dominant in the matter of balance-springs. In order to obtain isochronism in a balance-spring the essential feature to aim at is the coincidence of the centres of gravity of both balance and spring. When a balance is in perfect poise the centre of gravity is situated about the axis of rotation, which corresponds with the centre of the balance staff. With a perfectly fitted balance-spring one of the middle coils, known as the " dead " coil, moves partly in the path of the volute. The coils outward of the dead coil open and close uniformly, whilst those within, dependant more or less on the pinning-in angle in relation to the stud, show a correspondingly eccentric movement.

Normally the diameter of the spring is not less than the radius of the balance covering the outside of the rim and not more than that including the height of a screw and it may consist of from twelve to sixteen coils.

The renowned horologist, Abraham Louis Bréguet (1747–1823), originated the idea of bending the outside coil so that it was raised to a plane above the other coils. This practice, which has come to be known as the " Bréguet " overcoil, bore no reference at all to the theoretical terminal as it is now understood; it was simply a device to give freedom to the body of the spring. It was, however, noticed that with an overcoil under certain conditions the spring developed centrally whilst a flat spring threw over more on one side than the other, causing uneven friction on the balance staff pivots.

Still further it became apparent that a change of curvature in the overcoil produced also a change in the isochronism, and a watchmaker named Jacob, of Dieppe, was the means of inducing Professor M. Phillips, a French mining engineer, to conduct a mathematical research on this subject which lasted from 1858 to 1860. Professor Phillips' findings are carried into practice to-day, and it can be said that through them the whole art of timing was placed on a scientific footing. His *Mémoire sur Le Spiral Réglant des Chronomètres et des Montres* (1861) is a very exhaustive treatise and he demonstrates the advantages of various theoretical terminals which fulfil the requirements as to isochronism.

A later publication, *Manuel Pratique sur le Spiral Réglant des Chronomètres et des Montres*, forms a concise summary of the former treatise, and the conclusion at which he arrives is that the centre of gravity of the terminal must fall on the radius which completes the first quadrant of the spring through which the curve passes, and that it must be distant

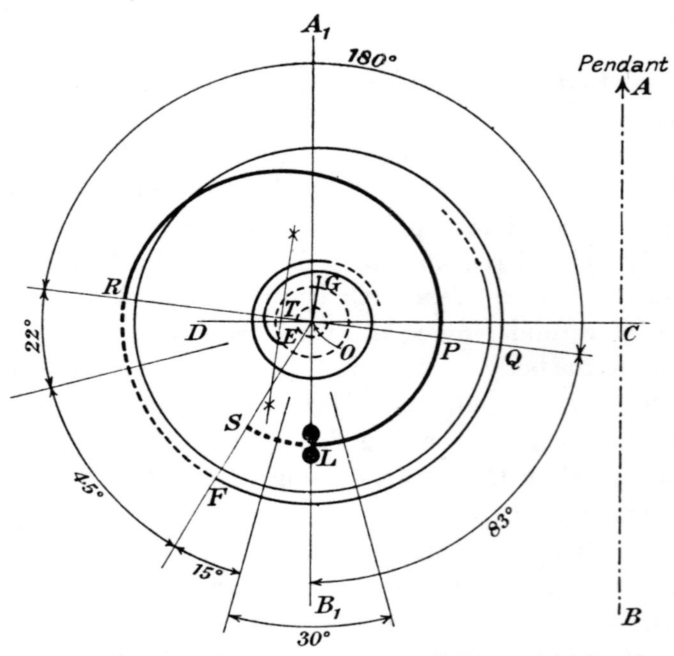

FIG. 72.—Illustrating a theoretical terminal designed by Lossier to Phillips' formula and frequently applied to modern watches.

from the centre of the spring by an amount which equals $\frac{R^2}{L}$; where R is the connecting radius and L the length of the theoretical terminal.

This formula represents the whole basis governing the construction of theoretical terminals.

Phillips shows numerous designs at the conclusion of his *Mémoire*, though some would not be convenient for practical use. To Jules Grossmann and his pupil, L. Lossier, credit is due, however, for expounding the theories of Phillips so that they could be more easily followed by the practical man. A form of theoretical terminal which is very often applied in present-day watches is shown in Fig. 72, and this is one of Lossier's designs.

The outside coil of the spring is shown by the thin circle whose radius is OQ, O being the centre of the spring. The "bend up" commences at F through an angle of 45° and continues in the same curve or volute for another 22° to point R. This is shown in heavy dots and OR is the radius of the outer coil. P on the diameter RQ is now marked off so that OP is 67/100ths of RO, and the bisection of RP gives the centre, T, of the next part of the curve, which completes a semicircle at P. From P, the curve resumes its course concentric to centre O. Passing through an arc of 83° to position L, the overcoil comes up to the central position of the curb pins, which are permitted to operate over 15° on either side. The effective length of the spring terminates at L, although the pinning-in takes place at the stud S. The length of the theoretical terminal is RPL and, by way of demonstration, the centre of gravity is shown at G.

If it be assumed that the connecting radius of the outer coil of the spring is 5 mm., then the distance $OP = 3{\cdot}35$ mm. The length of the curve RP is then :—

$$\pi\frac{(RO + OP)}{2} = \pi\frac{8{\cdot}35}{2}$$
$$= 4{\cdot}175\pi \text{ mm.}$$

$$\text{The length } PL = \frac{2\pi OP \times 83}{360}$$
$$= \frac{2\pi \times 3{\cdot}35 \times 83}{360}$$
$$= 1{\cdot}544\pi \text{ mm.,}$$

so that the total length of the theoretical terminal is :—

$$RP + PL = 4{\cdot}175\pi + 1{\cdot}544\pi$$
$$= 5{\cdot}719 \times 3{\cdot}1416$$
$$= 18 \text{ mm. (approx.)}.$$

To this must be added the amount required for the connecting piece, *LS*, which completes the overcoil in bringing it up to the stud.

The aim of the springer is to obtain accurate time with the index at the middle position on the scale; that is where

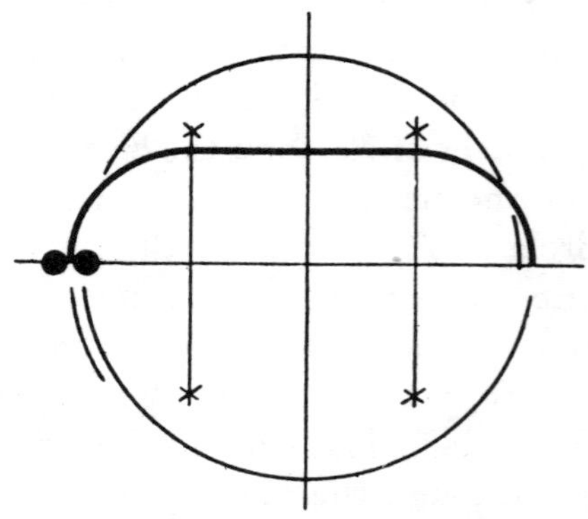

FIG. 73.—Illustrating another theoretical terminal often met with and evolved by Phillips.

the curb pins are shown in the figure. With the balance at rest, the outer coil of the spring, whether an overcoil or flat, should be midway between the pins, with the smallest possible clearance for freeing the blade. During the process of timing the index must remain on the line of centres. The fitting of a balance-spring with a theoretical terminal in accordance with the rules laid down involves quite a number of considerations which render close counting to time in the preliminary work superfluous. The usual procedure is reversed and by means of the balance screws the weight is altered either way until it conforms to the required number of beats.

Another theoretical terminal, evolved by Phillips and often met with, consisting of two quadrants connected by a straight line, is shown in Fig. 73.

An important feature to notice is that, where an index is used, it is necessary to choose a curve, part of which follows the arc of a circle concentric with the balance. So that, whereas many curves may solve the problem of isochronism, only a few lend themselves to practical application. It is essential that the sections of the curve merge into one another without any kinks.

Pinning-in at the Collet. There is another feature about the balance-spring which deserves close attention, that is the position of pinning-in the inner coil to the collet. Reference has been made to the "dead" coil which moves in the path of the volute. The coils outside the "dead" coil are under the influence of the overcoil, those within are not. It was discovered by Jules Grossmann that the common centre of gravity of these inner coils shifted away from the centre of the balance, and according to the arc described, showed itself by influencing the rates in the vertical positions from losing to neutrality and gaining.

Under normal circumstances there is a shortening of the arc of vibration of the balance when the watch is removed from the lying to the vertical position, and the two are generally, but erroneously, distinguished by the term "long" and "short" arcs respectively. It became apparent, therefore, that the effect of the eccentric centre of gravity of the spring could be rendered advantageous if, placed in a certain position and under the best mechanical conditions, it would show a slight gaining rate. The "pendant-down" position, being the most unlikely to occur in wear, could afford to be neglected, and watches are consequently never timed under such circumstances.

Returning to Fig. 72 and imagining the back of the

watch to be uppermost, the line A_1B_1, through the balance centre, O, is parallel to AB, the line of centres through the pendant. The line of centres, COD, of the balance-spring lies at right angles to A_1 B_1, and the first coil on leaving the collet at E must develop above the line of centres in order to fulfil the requirements of the "pendant-up" position. Accordingly it is the practice to ensure that the pinning-in takes place well below the line of centres, COD, and thus also to allow for shortening the spring if the need arises.

Normally the balance arcs are about 1¼ to 1¾ turns in the lying position, but when the watch is removed to the pendant-up position then the effect of the eccentric centre of gravity of the spring comes into operation. This manifests itself in a slightly gaining rate of about 1 second a day. It should be noticed, however, that if through defective conditions the arcs fall off to less than 1 turn, a reversal of the order takes place and the vibrations are actually retarded.

The extent of such an error may vary in amount to as much as 13 seconds a day, which will be found to exist under normal conditions if a watch is tried in the pendant-down position.

In order to achieve the best possible result in the pendant-up position, various devices have been introduced as a means of final adjustment. For example: a movable stud and index; a circular balance cock; an adjustable collet attachment.

Curb Pins. It has been shown that the effective length of the balance-spring terminates at the curb pins, and it is important that the blade of the spring should be only allowed the smallest amount of freedom between them. Too much play between the pins brings into operation the section of spring connecting the terminal with the stud, which thus increases the effective length and naturally produces a loss. This loss is greatest when the arc of vibration is least, so that a very distinct difference is apparent between the long and short arcs under these conditions. Careful adjustment of the pins is, however, necessary to counteract any centrifugal tendency in the balance. The acting diameter of the balance is increased in the long arcs through centrifugal force, particularly if the screws are accumulated near the cut and much depends upon the attainment of a rhythmical performance of the spring between the pins to neutralise the continuously changing conditions arising from the motion of the balance.

Inner Terminals. Returning again to the subject of the inner coils, there is a method of eliminating the existence of the eccentric centre of gravity with its consequent effect upon the timing in the vertical position. Bréguet endeavoured to overcome this difficulty by planting the entire escapement in a revolving carriage, which was a most delicate and costly apparatus. Bonniksen, Taylor, Lange, and others improved on this with a more substantial and slower-moving device, but all had the disadvantage of cramping the escapement to undesirable proportions and trouble often arose through oil finding its way on to the carriage, whose bearings were brass upon brass (cf. Chapter XIX).

It is nevertheless possible to overcome almost all position difficulties and yet use a properly proportioned fixed escapement by introducing an inner theoretical terminal to the balance-spring.

Fig. 74 shows the Lossier terminal applied in this way, and also the form of collet which is needed to meet the requirements of the curve. This collet has to be carefully constructed and poised. The effective length of the terminal commences directly the spring leaves the collet and ends on reaching the last point of the volute. The motion of the spring, when two theoretical terminals are used, is perfectly uniform and the inner coils lose entirely the uneven development which is ordinarily apparent.

On Plate VII is shown an application of this theory. The balance-spring is from one of Mr. Otto's adjusted watches, executed by him. The outer terminal is his own,

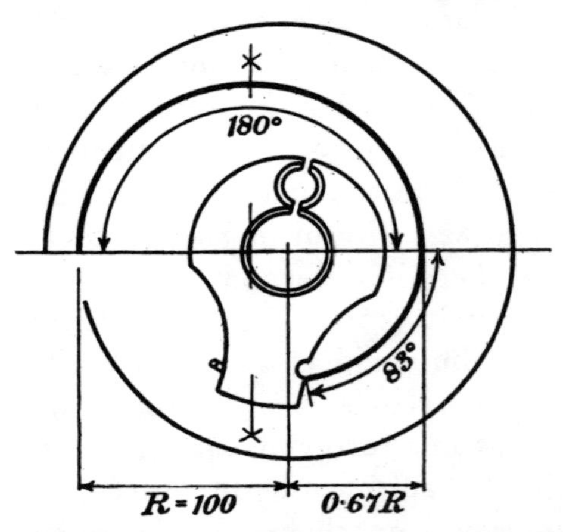

FIG. 74.—Illustrating the "Lossier" terminal of Fig. 72, applied as an inner terminal for the purpose of eliminating the eccentric centre of gravity with its deleterious effect upon timing in the vertical position.

(O), formation and has a connecting radius of 4 mm., the inner one is a Lossier, (L), with a connecting radius of 2 mm. For exact checking the colour has been removed on top of the blade of the inner terminal. The terminal O contains an undisturbed arc of the volute covering 60° (measured by using the length of the connecting radius), which is embodied in the construction. He gives details of this in his essays on timing, published in *The Watchmaker, Jeweler, Silversmith and Optician* (May 1911, page 591; March, April 1913, pages 317, 457).

Plate VIII shows this spring mounted on the balance.

PLATE VII.—The balance-spring of an adjusted watch by Mr. Otto, to which he has applied the inner and outer theoretical terminal.

PLATE VIII.—The balance-spring of Plate VII shown mounted on the balance.

Photographs by Mr. W. Beckmann.]

[*To face page 172.*

CHAPTER XIV

WATCH ESCAPEMENTS

WATCH escapements fall into a category similar to that for clocks, but the now obsolete verge, which is a small adaptation of the clock crown wheel, is the only recoil escapement. The remainder are all dead-beat, but subdivided as shown in Table V.

TABLE V.—WATCH ESCAPEMENTS

DEAD-BEAT	
Frictional-rest	Detached
Cylinder or Horizontal Duplex	Lever Chronometer

All the earliest watches had verge escapements and the first improvement on this which ever came into general use was the cylinder escapement. The idea was conceived by Thomas Tompion (1638–1713) about the year 1695; it was, however, George Graham, famous for the dead-beat clock escapement, who finally presented it as a practical proposition some five and twenty years later. Although this escapement showed distinct advantages over the verge, it is a curious fact that it did not revolutionise the design of watches in this, the country of its birth, but found considerably more favour on the Continent. There it soon became possible to produce inexpensively the somewhat peculiar escape wheel and cylinder in highly finished, hardened and tempered steel, which added considerably to durability.

The verge escapement was planted vertically in the movement, but the cylinder, being the first to be arranged horizontally, became also known as the horizontal escapement.

This escapement is fast becoming superseded, but it is certainly worthy of the estimable place it has held in the evolution of horology. Superiority of action, denoted by increased vibration and greater freedom for the balance, was the principal feature which raised it considerably above the level of the verge; whilst, combined with simplicity of construction, a well-proportioned escapement would give a very fair result in wear. This condition can only exist, however, if hard wheels and cylinders are used and if the balance and escape wheel bear the correct relationship with one another, that is, the radius of the balance should be equal to the diameter of the escape wheel. Failure of so many modern cylinder wrist watches is due to the fact that the escapement is of necessity cramped and the correct proportions are not maintained.

The Cylinder Escapement. The escape wheel teeth resemble a wedge in form, but are slightly curved. They are raised on pillars to a higher plane than the flat of the wheel, and the cylinder on which the balance and spring are mounted has a section of its shell removed to provide a passage for these escape wheel teeth to enter and leave, one at a time (see Fig. 75). The inclined outer surfaces of the teeth act as impulse planes to the balance by operating alternately on the two lips of the cylinder shell presented by the broken section. Locking is caused by the points of the teeth dropping on the shell, first on the outside and afterwards, as the balance turns it to the passage, on the inside. The period during which the point of the tooth impinges against the shell constitutes the frictional-rest.

Many early English cylinder escapements had a large

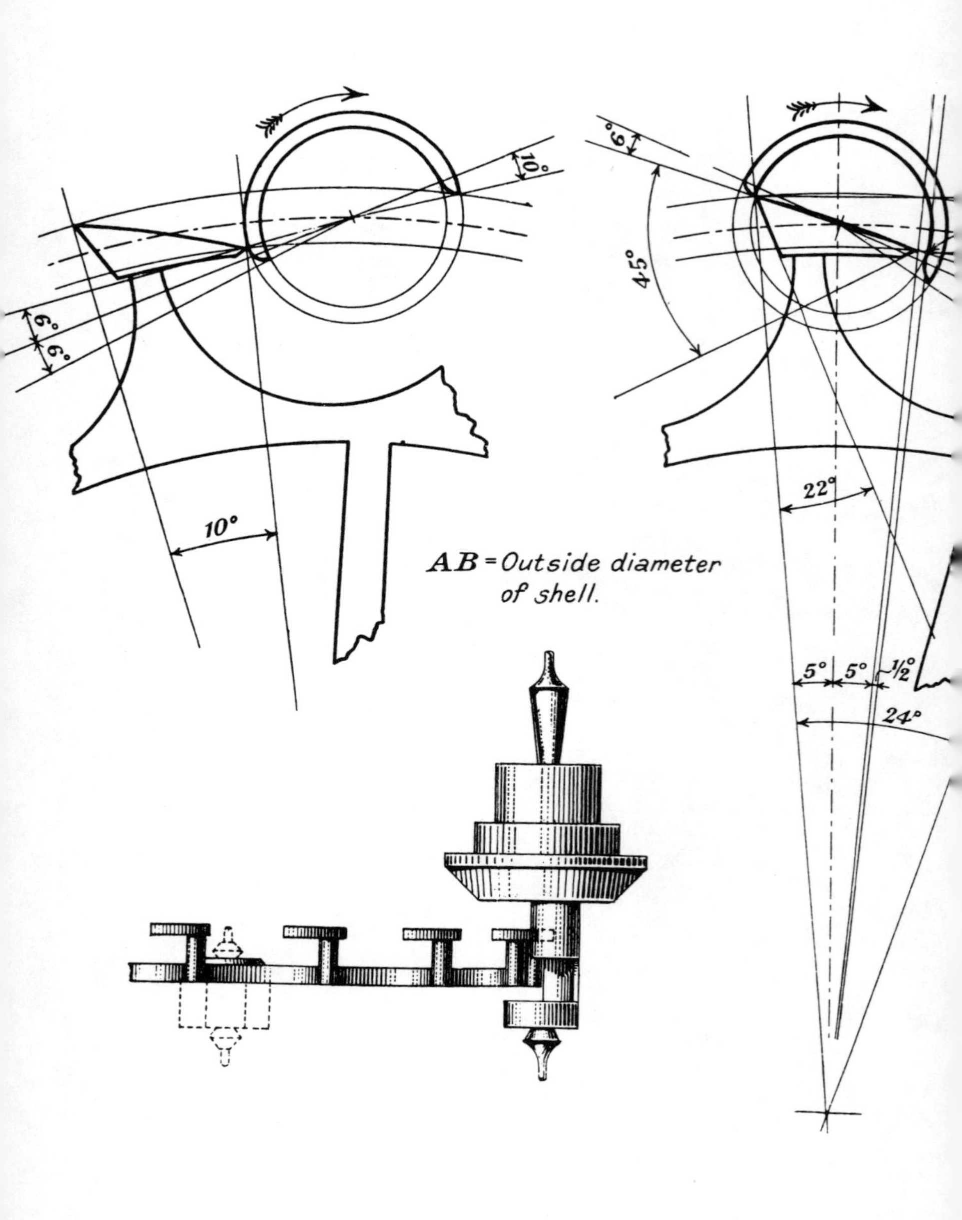
10°
6°
6°
10°
AB = Outside diameter of shell.
6°
45°
22°
5°
5°
1/2°
24°

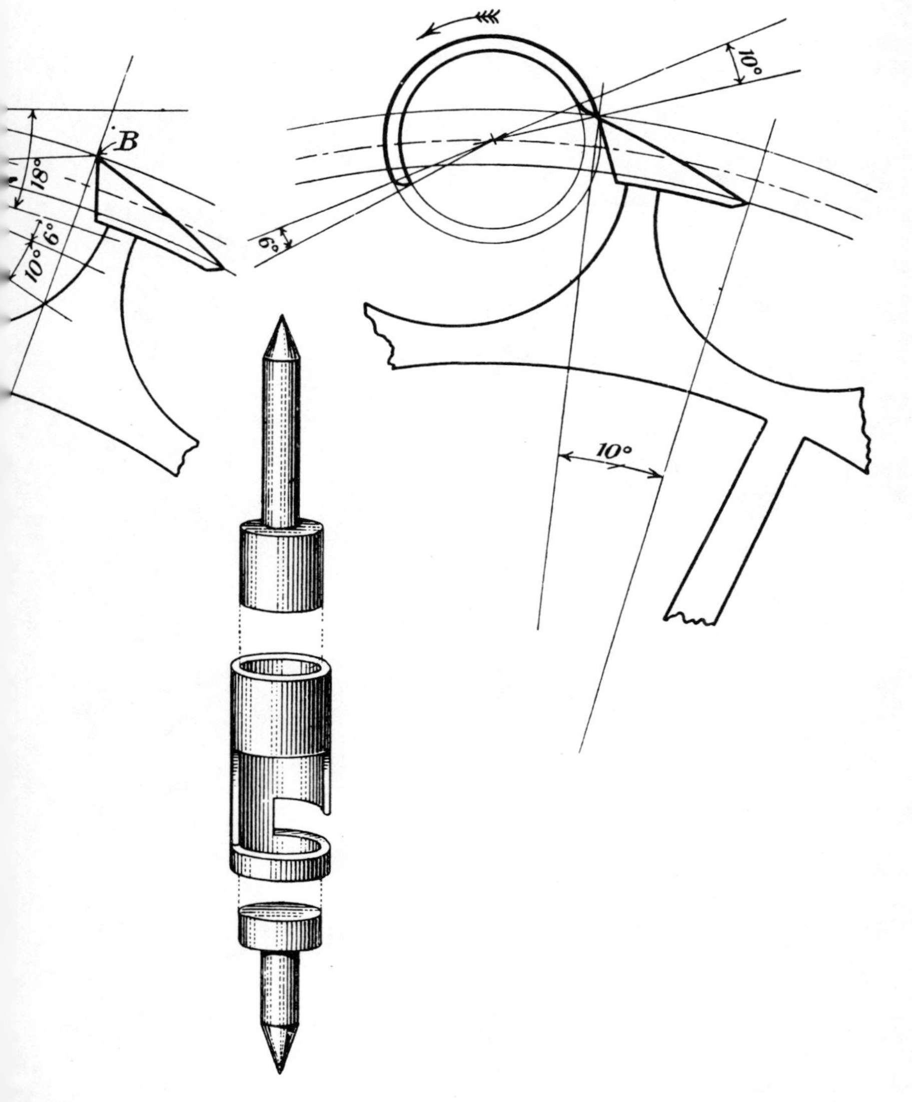

FIG. 75.—Cylinder escapement in its modern form of construction; originated by Thomas Tompion about the year 1695 and improved upon by George Graham. Inset left: The escapement in elevation. Inset right: The steel cylinder plugs and shell.

[*To face page* 175.

brass wheel and ruby cylinder, but the modern form, so extensively applied to Continental watches, has a light steel wheel and cylinder, both of which, as already stated, require to be hardened and tempered and highly polished. This form of escapement is illustrated in Fig. 75, the dimensions given being those generally accepted from experience as producing the best results.

The action is shown in three stages: First, with a tooth locked on the outer part of the shell of the cylinder; secondly, the same tooth entering the cylinder and having completed giving impulse, about to drop on the inside locking; and, thirdly, the tooth leaving the cylinder, impulse again just completed, with the heel of the tooth almost free.

The escape wheel, having fifteen teeth, gives an angle of 24° occupied by a tooth and a space. The length of a tooth from toe to heel covers an arc of 10° with the line of centres, and ½° is allowed for drop. The impulse plane of the tooth is inclined from the heel at an angle of 18° and slightly curved with an arc of equal radius to the wheel. In small watches, this angle is increased to 22°. The entering lip of the cylinder is sloped off at an angle of 6° with the impulse plane and the exit at 10°. An angle of 6° is allowed for locking both outside and inside the shell. The back of the tooth is undercut at 22° in order to give safe clearance, whilst the front is set back at 45° to the 18° line for the sake of durability. Both lips are suitably rounded so as to provide smoothness of action and correct transmission of impulse.

The Duplex Escapement. The duplex, which is another frictional-rest escapement, originated in France about the year 1724 and was invented by J. B. Dutertre. Although applied to many watches in different forms, it is an escapement which cannot be said to have achieved popularity at any time. This is doubtless attributable to certain constructional difficulties. First, it involves the use of an escape wheel with two sets of teeth which is extremely troublesome to cut, either individually or in quantities. Secondly, the balance staff requires to be very thin in order to take a small cylindrical ruby roller and is consequently very sensitive and liable to be easily broken. Thirdly, the action of the escape wheel upon the ruby roller is such that any fluctuation of force affects immediately the vibrations of the balance and produces errors in timing. Further, shallow intersection of the wheel and ruby roller may very readily render a liability to tripping, whilst too large a roller would introduce detrimental friction. Fourthly, and what may be regarded as the principal disadvantage, it is a single-beat escapement, so that for pocket wear a sudden movement of the body under certain circumstances would be sufficient to cause a set and the watch would stop.

Nevertheless, many fine duplex watches have been constructed, and the name of McCabe, the eminent manufacturer who practised in the early part of the nineteenth century, will always remain prominent by reason of the successful results which he achieved with them. His successes depended, however, on a strict and careful proportioning of the parts, especially the ruby roller.

The action of the duplex escapement is shown in Fig. 76. The balance staff carries a very small ruby roller, *R*, which is a cylinder with a narrow section cut out. Above this, also mounted on the staff, is an impulse pallet, *M*. The wheel, *W*, which has fifteen long triangular-shaped teeth, has a corresponding number of short teeth between the others projecting from the face of the wheel. The long teeth lock on the ruby roller and provide a portion of the impulse, but the short teeth are impulse teeth only and transmit impulse to the balance through the impulse pallet. Impulse is only given at alternate vibrations and, when

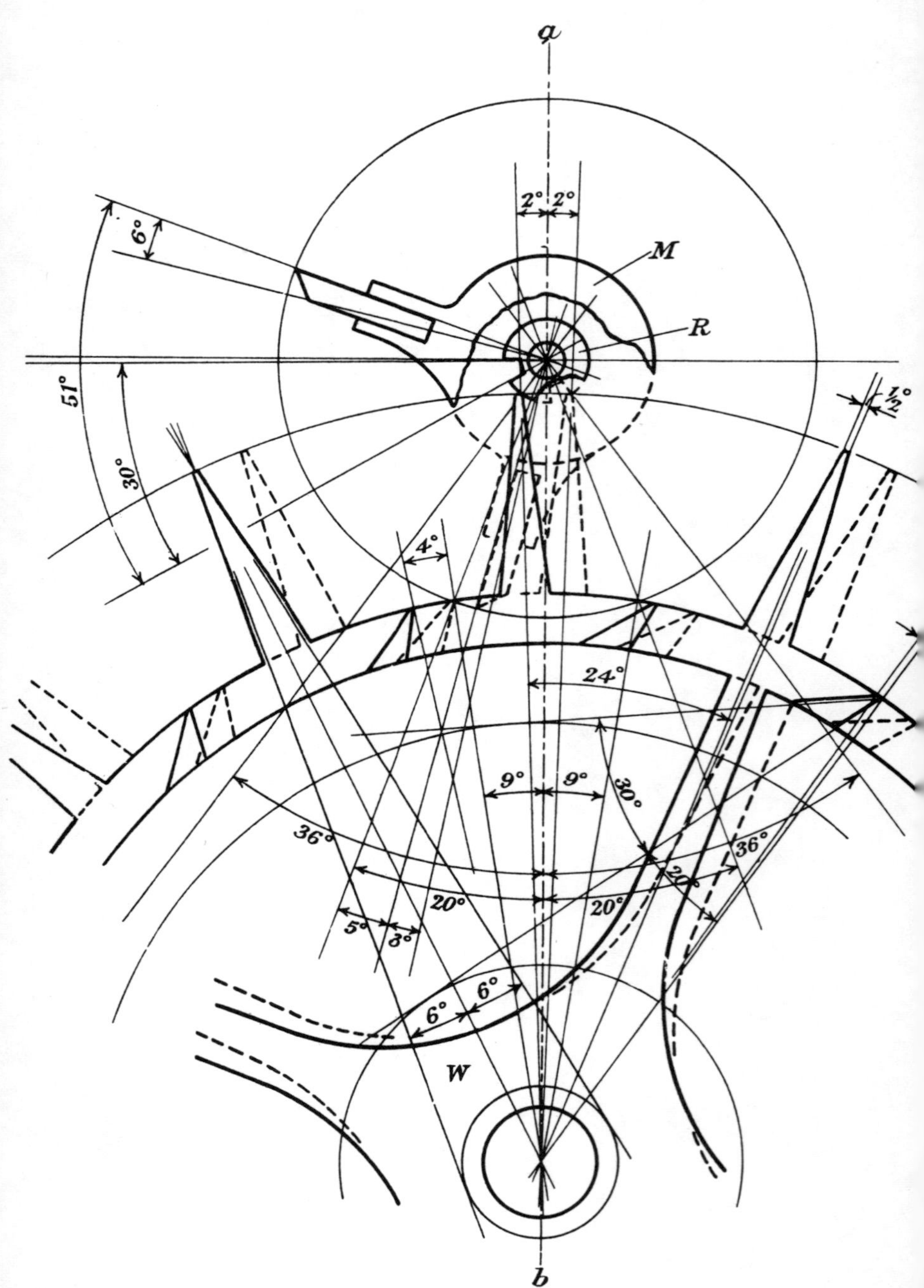

Fig. 76.—" Duplex " escapement. The original form was invented by J. B. Dutertre abou[t] the year 1724.

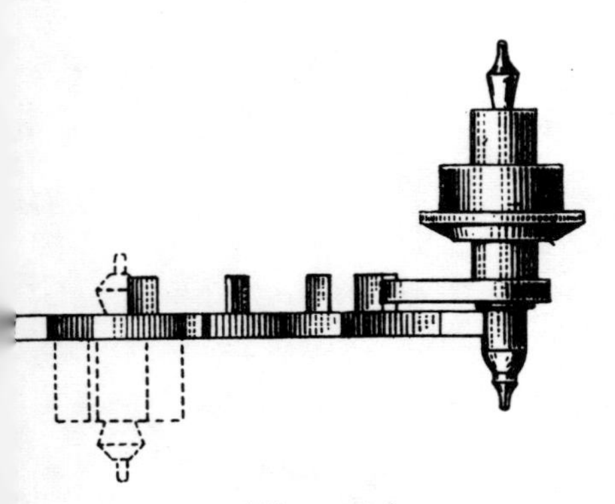

Elevation

[To face page 177.

this has taken place, the next long or locking tooth drops on the ruby roller, the balance makes its return vibration and carries the ruby-roller notch through an equal angle past the point of the locked tooth. During the succeeding vibration to the right, the tooth enters the roller notch, imparts some of the total impulse, and so precedes the subsequent advance of the next impulse tooth. This in due course drops on to the impulse pallet and the entire impulse is completed. There is proportionate friction between the escape wheel and locking faces both of the ruby roller and impulse pallet.

Although this escapement is obsolete it is not infrequently met with. Makers had different ideas on the subject of suitable proportions, but the dimensions given in the figure are applicable to the majority of cases and would ensure correct working.

Two successive stages in the action of the escapement are shown. First, a relative position of the wheel, the ruby roller and the impulse pallet at the moment when the balance has completed its vibration to the left and is about to return. The wheel has a tooth locking on the ruby roller and impulse will commence as soon as the notch admits the tooth. Secondly, in dotted outline, the same locking tooth just being released by the roller. The impulse pallet has travelled round with the roller to a position of 5° in advance of the next impulse tooth. As soon as the locking tooth moves clear of the roller the next impulse tooth drops through this angle of 5° on to the pallet and proceeds to transmit the remainder of the impulse.

The intersection of the ruby roller with the locking teeth of the escape wheel is 4° (2° being measured on either side of the line of centres, *ab*), making an impulse angle of 72° with the roller. The impulse pallet intersects the impulse teeth through an angle of 9° on either side of the line of centres, *ab*, but 5° are allowed for drop and 3° for clearance, so that, although this intersection makes a total angle of 40° with the roller, the actual angle of impulse is only 32°. During alternate beats, therefore, the balance receives 72° of impulse from the locking tooth and 32° from the impulse tooth, making the complete angle of impulse 104°. The angular distance between the locking face of the ruby roller and that of the impulse pallet is relative and in practice permits of slight adjustment. In the drawing they are separated by an angle of 51°, that is, $36° + 20° = 56°$, representing the total intersection up to the line of centres, less the 5° which are allowed for drop. 6° covers the width of the impulse pallet.

The flanks of the locking teeth are tapered from the tips at an angle of 6° on either side of the radius, whilst the impulse teeth are undercut 20° with backs inclined at 30°. In both cases $\frac{1}{2}°$ is allowed for the tops of the teeth.

Lever Escapements. Whatever merits may be claimed for frictional-rest escapements, one thing is clear, that, as to time-keeping properties, they cannot attain to those of detached escapements. This is by no means surprising, as one of the essential features of accurate performance is that the balance and spring should possess the greatest possible freedom of motion, unhampered by any interference arising from deficient construction of the train. The lever escapement, as it is to-day, approaches this desirable condition to a remarkable degree. Within the limits possible in actual practice it is hard to imagine anything so closely realising the theoretical purpose. The principle is accredited to the eminent horologist Thomas Mudge (1715–1794). Nevertheless, it was the result of numerous applications of his ideas by other watchmakers that the modern designs and proportions of the escapement have

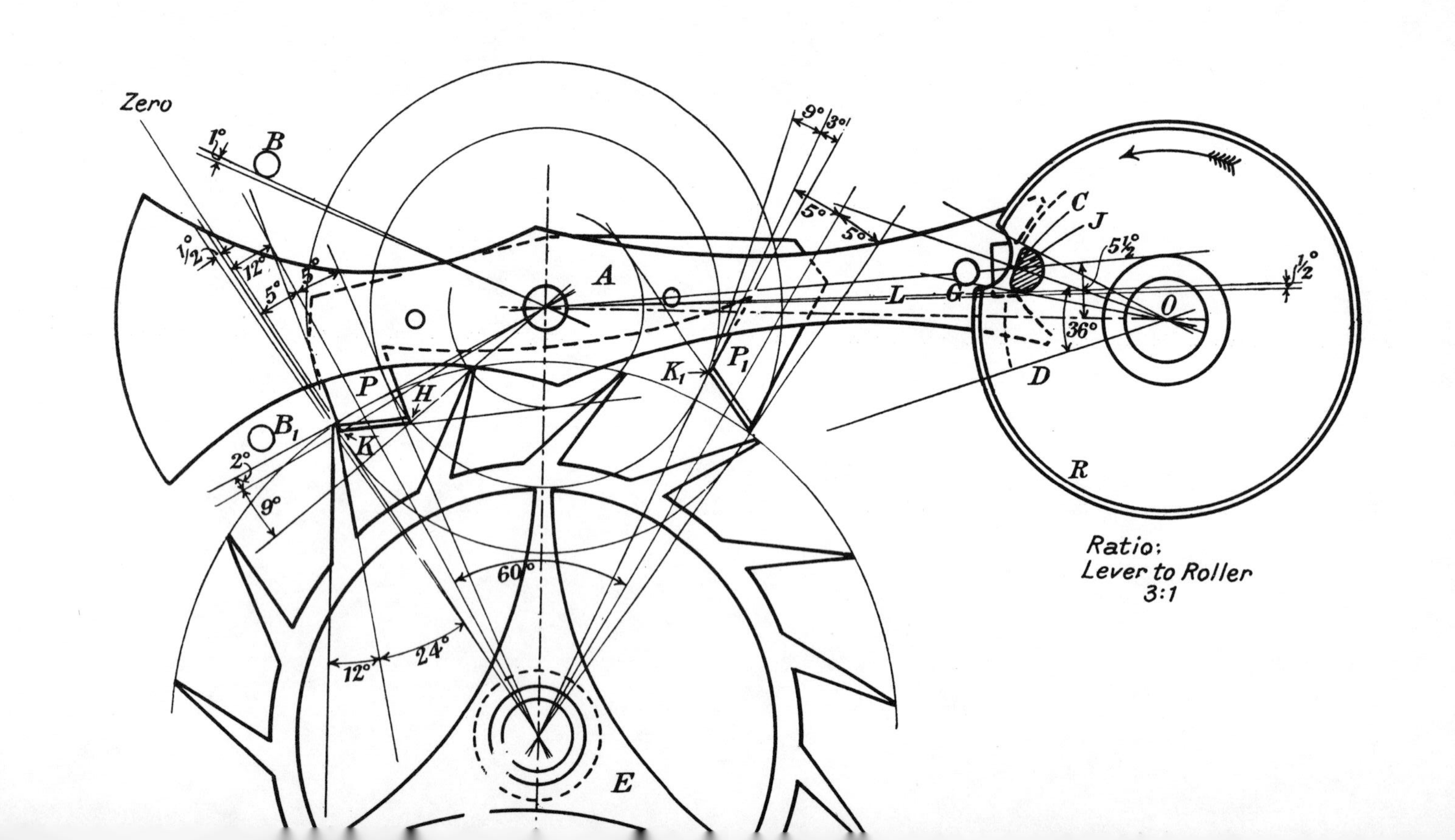
Zero
Ratio:
Lever to Roller
3:1

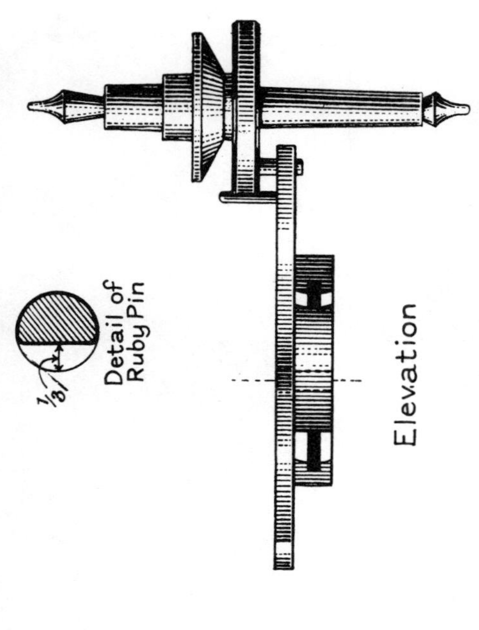

77.—Ratchet-tooth lever escapement. Showing the single roller form with circular pallets.

[*To face page* 179.

been evolved, which have so conclusively proved its worth. There are two principal forms of lever escapements, the one generally adopted by English makers of hand-made watches, the other having become almost exclusively adopted by foreign manufacturers. A third form sometimes appears in very common mass-production watches and alarum clocks, to which allusion will be made later.

The Ratchet-tooth Lever Escapement. The English form is shown in Fig. 77 and it will be noticed that the action is divided between the escape wheel, *E*, the pallets, P, P_1, the lever, *L*, and the roller, *R*.

The escape wheel has fifteen ratchet-shaped teeth acting alternately on the pallets, P, P_1. The wheel, *E*, advancing in the direction of the arrow is shown with a tooth at rest on the entrance pallet locking. The pallets, P, P_1, are mounted on the pallet staff and pivoted to move about the centre, *A*. Attached to the pallets is the lever, *L*, which operates usually, and in this instance, at right angles to the pallets. One end of the lever is forked and intersects the path of the roller, *R*, which is mounted on the balance staff and the roller carries the ruby-pin, *J*. The distance of centres between the pallet staff and the balance staff is divided so that three-quarters is allowed for the acting length of the lever and one-quarter for the acting radius of the roller. This acting radius extends only to the diameter of the ruby-pin, which in this instance is flattened in front, as explained later. Projecting vertically below the roller, the ruby-pin, following the motion of the balance, enters a notch which is part of the lever fork. In the figure the roller, moving in the direction of the arrow, has advanced the ruby-pin to a position where the action upon the lever has just commenced. The lever is removed from the banking pin, B_1, and has overcome the important angle for draw. The tooth on the entrance locking will be released next and impulse communicated to the balance by the action of the tooth on the pallet and the consequent propelling influence of the lever notch upon the ruby-pin. As soon as impulse is completed, the exit or discharging pallet is in readiness to receive the next tooth on the locking. One of the most important functions of the lever escapement follows the reception of the tooth on the locking. The teeth are undercut and the locking faces of the pallets inclined in such a way that the pressure of a tooth on the locking produces a drawing-in motion of the pallet towards the wheel, the extent of which is determined by the position of banking pins, B, B_1. "Draw," as this motion is called, is a detail of the greatest importance and absolutely essential to the correct performance of the escapement. By this means, the lever is brought up to the banking pins in order to ensure perfect freedom of the roller as the balance completes its vibration.

Occasions arise where outside influences overcoming the action of the draw produce a tendency for the lever to leave its position of rest on the banking pins prematurely, and a safety device is, therefore, introduced. This takes the form of a brass guard-pin, *G*, inserted in the lever near the notch, so that it projects upwards into the plane of the roller. A "crescent" or "passing hollow" is cut out of the edge of the roller on the line of centres of the ruby-pin. As the lever crosses the roller the guard-pin passes within this crescent but in the event of the lever leaving the bankings before the ruby-pin reaches the notch, the guard-pin butts against the edge of the roller and is thereby prevented from passing.

This form of safety device applies to what are known as single-roller escapements, where the balance arc is normally about 36° as shown in the figure. Should the usual relative ratio of 3 : 1 between the length of the lever and size of the

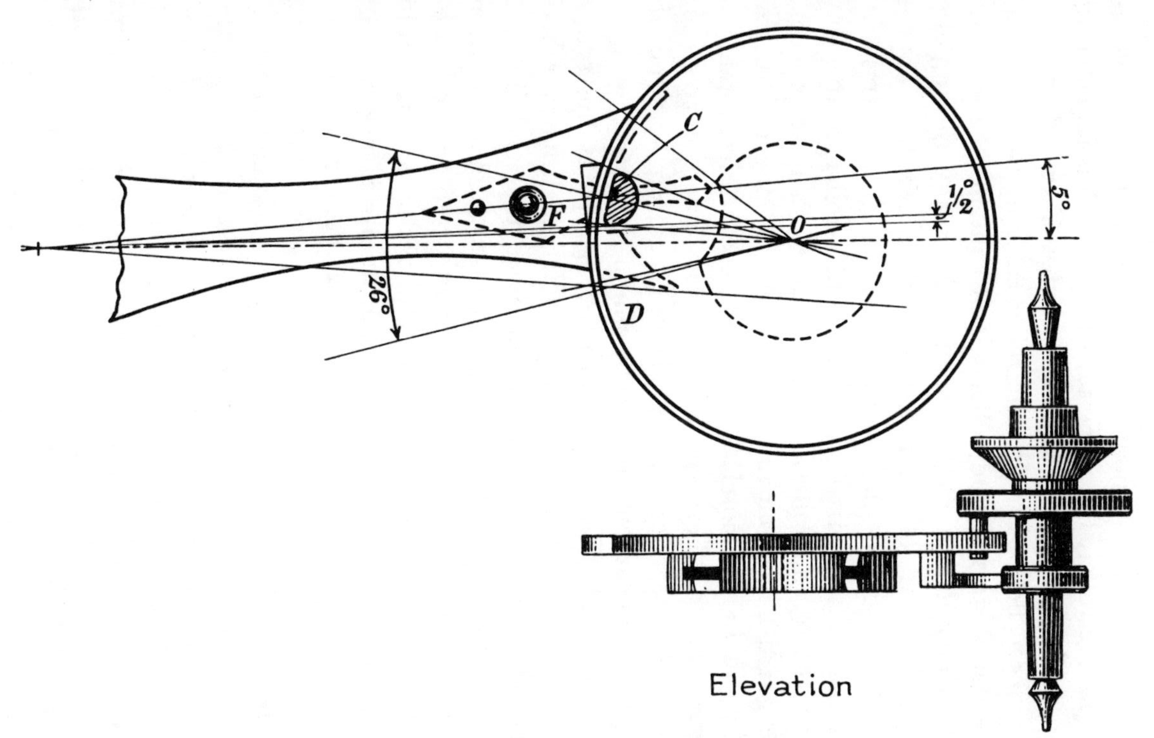

Fig. 78.—Double roller applied to the ratchet-tooth lever escapement.

roller be altered, and the balance arc be reduced below 33°, this device becomes unsafe, and another system, known as the double roller, has to be substituted. This is shown in Fig. 78, where it will be observed that a second roller is introduced below the impulse roller carrying the ruby-pin. The size of this second roller which contains the crescent is arbitrary, but it may be about half that of the impulse roller. Instead of a guard-pin, the lever carries a projection, *F*, called the guard-finger, which performs the function of the guard-pin of the single-roller escapement upon the second roller. In the double-roller escapement, the guard-finger remains in the crescent longer than the ruby-pin in the notch, so that it becomes important to ensure that the horns of the lever, the inner curvature of which is concentric with the balance centre, are long enough to permit safe action of the ruby-pin in the notch.

Returning to Fig. 77, which may be regarded as illustrating the typical proportions for a single-roller English or ratchet-toothed escapement, it will be observed that the different components are shown relatively in an artificial position whilst in motion. The figure is constructed on the line of centres with regard to the lever and roller, and the radii of these two, as previously mentioned, are in the ratio of 3 : 1. This combination gives a balance arc, *COD*, of 36° and a lever arc of 11°, the effective point of intersection being the centre section of the ruby-pin through point *C*.

In the case of the double roller shown in Fig. 78, the balance arc, *COD*, is only 26° and the lever arc 10°, so that the balance is thus given more free run.

The escape wheel has fifteen teeth, and these are undercut 24° to give clearance for draw. The width of a tooth occupies 12°, whilst ½° is allowed for the top. The pallets cover 2½ teeth, which represents an angle of 60° and forms the centre section of the acting surfaces of the impulse faces. Of the remaining half tooth, amounting to an angle of 12°, the pallet faces cover 10°, being 5° on each side of the angle of 60° and 2° to allow for clearance and drop. The pallet staff centre is at the intersection of chords through the pallet face angles of 10°.

The angle of motion of the pallets corresponding to the lever arc already mentioned is 11° and this is divided so that 2° are allowed for locking and 9° for impulse. It then follows that the impulse faces of both pallets must be tangents to the common circle known as the "impulse tangent circle." This circle is derived from a line which commences at the corner, *K*, of the entrance pallet, passes through the angle of 10° and continues until it reaches the pallet-path circle at *H*. Projected still farther, this line determines the radius of the impulse tangent circle, and the impulse plane of the exit pallet is also necessarily a tangent to this circle. The term "circular pallets" distinguishes this form from another described as pallets with "equidistant lockings," where the locking takes place at equal distances from the pallet centre, with the result that one pallet arm is longer than the other. These escapements are, however, only met with experimentally in English work.

Draw is effected by inclining the locking surfaces of the pallets 12° from the tangents to the pallet-path circles at the locking corners, K, K_1. In the figure the action is taking place at the entrance pallet, but there is a similar amount for the exit pallet at the moment of acting there. 1° is provided for run to the bankings. It will be noticed that as the angle for draw is taken from the locking corner of the pallet and the actual locking takes place 2° above, the tip of the tooth passes through the 12° before reaching the pallet and prior to the action of draw taking place. This position of the tooth is denoted by the line marked

" Zero " and forms the starting point for spacing off the other teeth in the wheel.

If the pallets are jewelled, the stones are set in longitudinal channels flush with the steelwork and rounded off crosswise so that only the smallest surface is presented to the wheel teeth. This is indicated by a double line.

The actual form of the ruby-pin in this escapement is usually " flatted." This is a cylindrical stone faced to a flat surface at a depth of one-third of the diameter, with the sharp edges slightly rounded off. The flat part of the pin is presented to the notch and affords good clearance in entering and parting from the lever. Other forms may be either oval or triangular.

The Club-tooth Lever Escapement. At the present time the club-tooth lever escapement is the most extensively used of all watch escapements ever invented. Though, in principle, it is similar to the ratchet-tooth escapement, it possesses certain differences which may be claimed as very distinct advantages. These differences mainly occur through the shape of the wheel teeth, which are shown in Fig. 79. Teeth of this form are stronger than the pointed ratchet teeth. The undercut at the heel also reduces considerably the amount necessary for drop, and bevelling of the edges is another feature which minimises contact with the pallet faces and enables the oil to be better retained.

These are facts upon which too much emphasis could not be laid, especially the matter of retaining oil, as they are the refinements which have raised the lever escapement to the high degree of efficiency that it now possesses.

In contrast to the design of the ratchet-tooth form in English watches, where the lever operates at right angles to the pallets, it is almost invariably the custom in foreign watches to plant the balance staff, pallet staff and escape wheel all on the line of centres. This arrangement is,

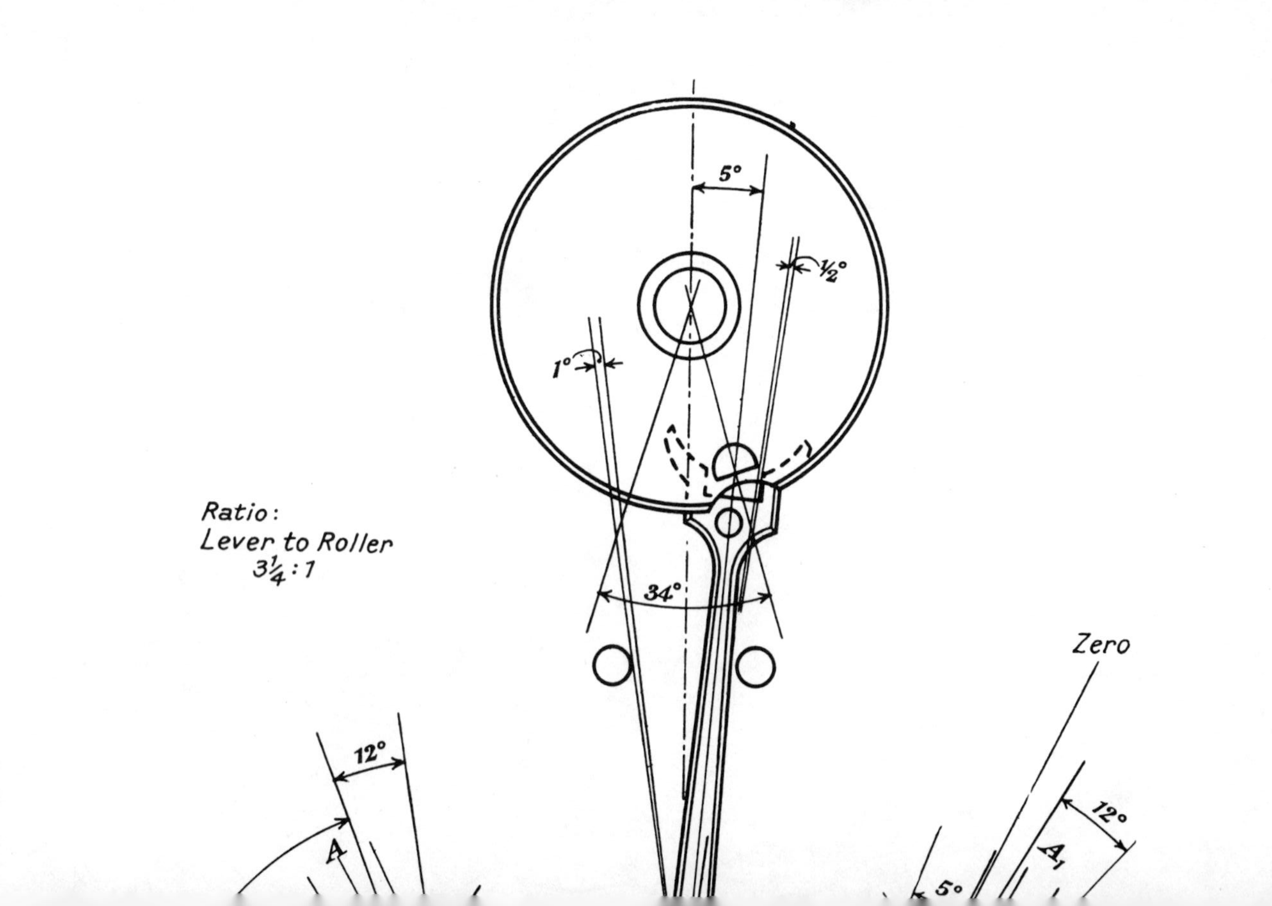
5°
½°
1°
Ratio:
Lever to Roller
3¼ : 1
34°
Zero
12°
A
12°
A1
5°

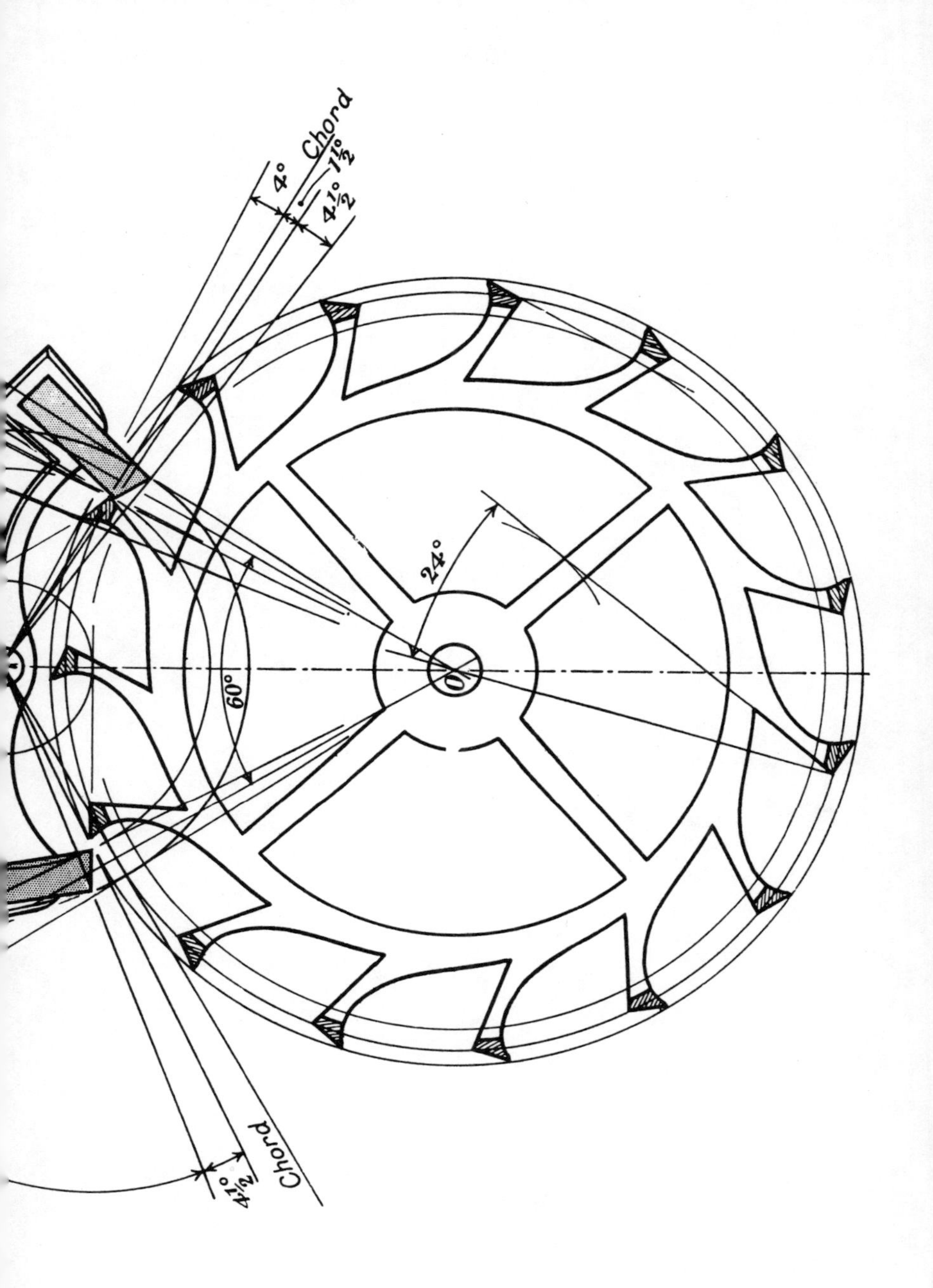

FIG. 79.—Club-tooth lever escapement with circular pallets.

[*To face page* 185.

therefore, often described as the "straight-line lever" escapement.

Fig. 79 shows the single roller form of this escapement and pallets with circular locking. The lever and roller proportion is slightly less at 3¾ : 1 than in the ratchet-tooth escapement. A tooth of the wheel is on the exit locking and the tops of the teeth are nearly as long as the pallet faces, so that, impulse is thus continued after the front tip of the tooth is clear. As in the former case, the wheel teeth are undercut 24° to give clearance for draw, but the tops cover 5°. The pallets embrace, again, an angle of 60°, AOA_1, equal to the space of 2½ teeth. The pallet impulse faces cover 6°, being 3° on each side of the 60° lines, OA, OA_1, and the pallet staff centre is situated at the intersection of the chords of these respective angles of 6°. These angles of 6° for the pallet impulse faces, added to the 5° for the top of a tooth, give a total tooth and pallet angle of 11°. 1° only is needed in this escapement for drop, because of the hollow at the back of the tooth. The lever arc of 10° is distributed from the pallet centre as follows: 4° for the inclination of the tooth, 1½° for locking and 4½° for impulse. The impulse faces, as before, must form tangents to the "impulse tangent circle," as shown in the figure. A line passing through each locking corner at right angles to that connecting them with the pallet staff centre gives the position for marking back the angles of 12° to provide the important inclination for draw. The wheel spacing again commences on the "zero" line where the tooth actually locks.

Fig. 80*a* illustrates an escapement with "equidistant lockings," that is to say the locking corners are at an equal distance from the pallet centre. The arm of the exit pallet is longer than that of the entrance, which makes it necessary to form them carefully in order to ensure perfect

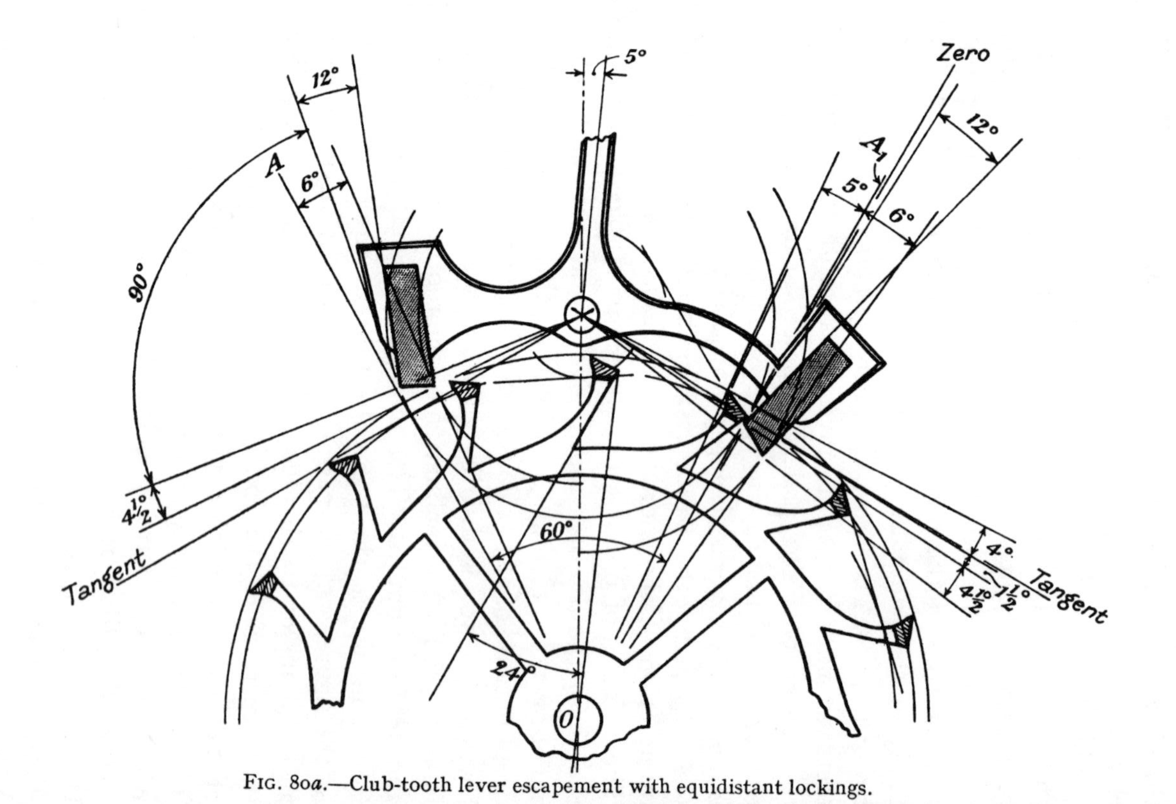

FIG. 80*a*.—Club-tooth lever escapement with equidistant lockings.

poise. The dimensions given in the figure will make it possible to compare this design with that of Fig. 79. It will be noticed that the impulse faces are no longer tangents to a common circle. The pallet staff centre in this case is situated at the intersection of tangents to the 60° lines OA and OA_1.

Whilst escapements with circular pallets are not often met with and those with equidistant lockings are reserved for highest grades, the vast majority of watches are now fitted with escapements which may be regarded as a compromise between the two. These are known as having "semi-equidistant lockings." Shown in Fig. 80*b*, it will be seen that the action is not strictly semi-equidistant but the distribution of the 6° allowance for pallet width is one-third and two-thirds on each side of the 60° lines OA, OA_1; that is 2° on the left and 4° on the right. 4° only is allowed for the top of the tooth so that there is a total tooth and pallet angle of 10° instead of 11°. From the pallet centre the lever arc of 10° is distributed rather differently: 3½° for the inclination of the tooth, 1½° for locking and 5° for impulse. By this method of distribution greater latitude is possible for obtaining a good action with less exactness of construction.

Pin-pallet Lever Escapement. This is a third form of lever escapement, but is met with only in very common watches and alarum clocks. Impulse amounting to about 2° or 3° is provided by the inclination of the wheel teeth acting upon pallets which are ordinary steel pins. A lever arc of 11° is usual and this is distributed between the pallet pin and the wheel to the extent of 2½° and 8½° respectively. Although many clocks and watches that are constructed with this escapement are capable of limited accuracy, the fact that draw has to be provided on both pallets by undercutting the wheel teeth to an angle of

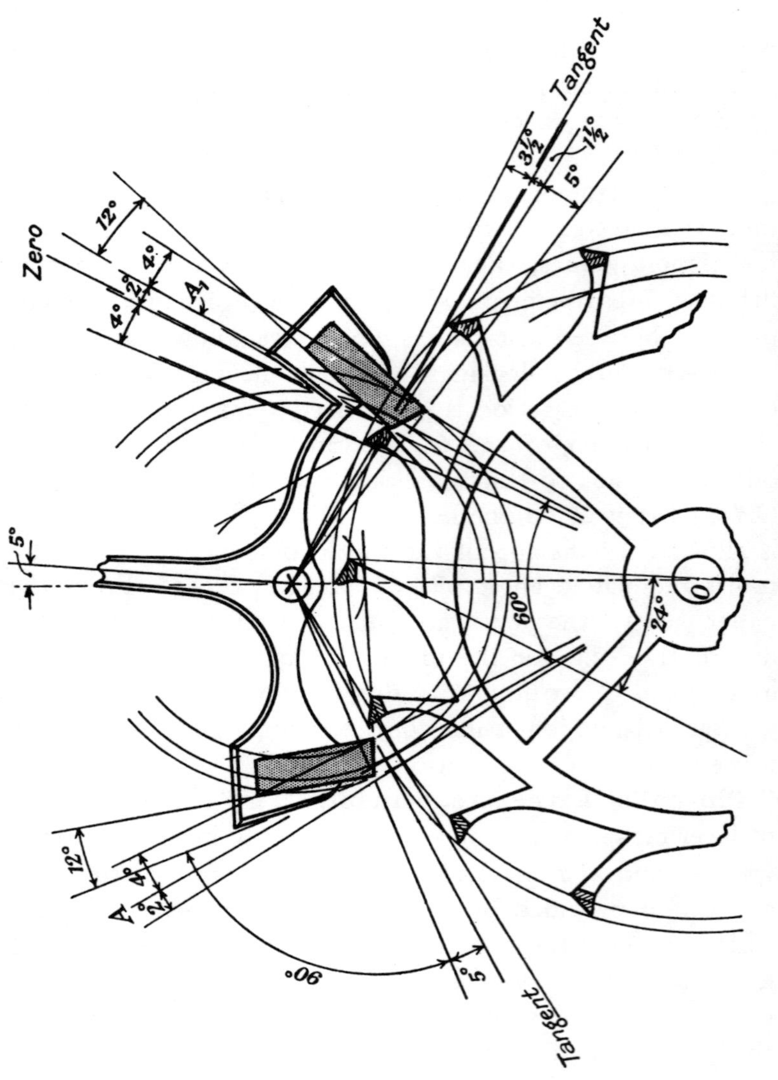

FIG. 80*b*.—Club-tooth lever escapement with semi-equidistant lockings.

$18°$ causes a bad tendency to set on the exit locking where the draw becomes excessive.

The other type of detached escapement, namely, the chronometer, is now never applied to pocket watches. It is, however, exclusively used in marine box chronometers, and a detailed description is therefore given in Chapter XX, dealing with these instruments.

This escapement, as applied to watches, has the same disadvantage as the duplex, inasmuch as it receives impulse only at alternate beats. The lost beat is always liable to cause the escapement to set, and damage can be done so easily to the delicate spring known as the "detent."

In Continental watches, this detent is sometimes pivoted, whilst, mounted on the arbor, is either a small balance-spring or some other form of return spring to bring it to the banking after the unlocking has taken place.

CHAPTER XV

MAINSPRINGS, FUSEES AND GOING BARRELS

Mainsprings. The consideration of mainsprings for use in watches involves to some extent reiteration of what has been said already, in Chapter VII, on the subject as applied to clocks. Mainsprings are divided into two main categories: those known as "lever," which are suitable for the old type of English fusee lever watches; and those known as "going barrel," "Geneva" and "Lépine," which terms apply equally nowadays to all forms of watches either with lever or cylinder escapements provided with going barrels. The contrast between the two, in the springs themselves, is that the one required for lever watches with fusees has to be strong and give few turns, whilst the going-barrel spring is longer and weaker. The grades of steel and suitable methods of hardening and tempering to satisfy these requirements are matters to which manufacturers have to give careful consideration.

It will be recalled that, in dealing with the subject of balance-springs, reference was made to the fact that in the coiling of a spring—one end being rigidly fixed—the bending moment is directly proportional to the angle of winding. That is to say, if a spring wound through n turns offers a resistance of x grammes, then on winding it through $2n$ turns the resistance offered will be $2x$ grammes, and so on. This feature applies also to a mainspring, but whereas the balance-spring is wound from rest, the mainspring is already partly coiled when inserted in the barrel before the watch is assembled.

Fusees. The barrel of the fusee watch is required to make about four turns and it is usual for the spring to have eleven coils when lying in the barrel. Out of the barrel it has approximately five coils, so that in winding it into the barrel the spring will have completed

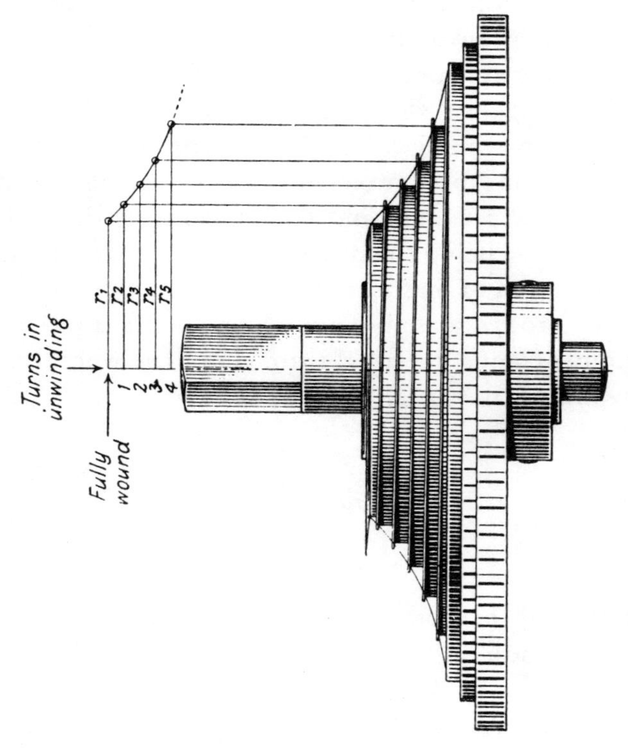

FIG. 81.—Illustrating the curvature of a fusee.

(11 − 5) = 6 turns of winding from rest. The first turn of winding in the barrel would thus become the seventh turn of winding of the spring. Assuming a spring out of the barrel on completion of the first turn to exert a force of x grammes, by the time it is lying in the barrel it is exerting a force of $6x$ grammes, then the four complete turns of

winding in the barrel give respectively forces of $7x$, $8x$, $9x$ and $10x$ grammes. If, therefore, the force at the fusee edge has to be maintained constant, then the moment of that force; that is the force exerted by the spring multiplied by the radius of the fusee, or strictly speaking, the distance of the edge from the axis of rotation, has to be the same at all stages of unwinding.

The requisite curvature for the fusee can be plotted graphically against the turns of winding and the varying radii, as shown in Fig. 81. Taking the case already given, the maximum force exerted by the spring when fully wound is $10x$ grammes and the moment at the fusee edge is $10x \times r_1$ (r_1 being the distance from the axis of rotation). On completion of one turn of unwinding, the force exerted is reduced to $9x$ grammes and the moment is $9x \times r_2$, but the moments must be equal throughout the whole operation of unwinding, so that :—

$$9xr_2 = 10xr_1$$

whence

$$r_2 = \frac{10}{9}r_1$$

$$= 1{\cdot}111r_1 \text{ (approx.)}.$$

Similarly, at the end of the 2nd, 3rd and 4th turns the respective radii are found to be 1·25, 1·428 and 1·666 times r_1.

The steepness of the curve depends upon the space available for the fusee between the frames. Thus the curve of a watch fusee is more pronounced than that of a clock fusee of 16 turns where the barrel makes 8 turns.

The number of turns made by the fusee depends on the design of the train. A watch is arranged to go for 30 hours, so that, if the fusee is to make 4 turns as well as the barrel, the great wheel must have 75 teeth and the centre pinion 10 leaves.

Expressed as a formula :—

$$\frac{GT}{c} = 30$$

where G = number of teeth in the great wheel,
c = number of leaves in the centre pinion, and
T = number of turns of fusee.

Thus, $$\frac{75 \times 4}{10} = 30.$$

This is quite a customary combination. If, however, a wheel of 72 is used with a pinion of 12, the equation becomes :—

$$\frac{72T}{12} = 30$$

whence $$T = \frac{12 \times 30}{72} = 5,$$

so that the fusee then makes 5 turns.

The Adjusting Rod. In practice, the evenness of the "pull of the fusee" may be tested by means of an "adjusting rod." This is a steel rod with brass jaws at one end to clamp it to the fusee arbor and adjustable sliding weights at intervals along the rod. The plates, with the fusee and barrel in position, are held so that the rod when attached to the fusee arbor can move in a vertical plane. The mainspring is set up and the chain partly wound on to the fusee by means of the adjusting rod. The weights on the rod are then shifted about until the pull in unwinding is sufficient to turn the rod gently : in other words, when the turning moment of the rod balances the turning moment of the fusee. The next operation is to wind the chain completely on to the fusee, and during the process of unwinding observe carefully the fluctuations of the pull at all stages. If the fusee is correctly cut the only adjustment which can be made is by altering the "setting-up" of the mainspring or by changing, if necessary, the mainspring itself. Under special circumstances a mainspring can be constructed in such a way as to be weaker in places.

If the spring is set-up too much the pull on the adjusting rod will be greater as unwinding proceeds, whilst it will become feeble towards the end if the spring is not set-up enough. A badly cut fusee generally shows too much pull in the middle turns.

Maintaining powers similar to those described for use in clocks are invariably used in watch fusees and need no further description.

Going Barrels.—The fusee is no longer a feature in watches, because the construction of escapements has been so much improved that they do not allow the timekeeping to be seriously affected by variation in the pull of the spring. In going barrels, the mainspring must be weaker than in the case of a fusee and contain about 13 coils after being wound into the barrel. This increased length and reduced strength of the spring also contribute towards reducing to some extent the variations of force between the consecutive turns. In high-grade watches it is possible to help matters still further by means of a stop-work, similar to that described for use in clocks, in order to employ only the middle turns of the spring within the barrel. The barrel is, in such cases, almost invariably arranged to make four turns. An enormous number of watches, in fact the majority, are now made, however, without stop-works, which have to be carefully fitted, otherwise they are likely to cause trouble. If a 5-turn stop-work is used, the spring makes $6\frac{1}{2}$ turns, $\frac{3}{4}$ of a turn being set-up.

Fig. 82 illustrates a high-grade form of stop-work in

which the star wheel rides on a brass pipe in the plate instead of being held down by a shoulder screw. Allowing sufficient freedom, a large-headed screw is screwed into the pipe to keep the star wheel in its place. One of the butting corners of the finger is suitably rounded to give a specially smooth and safe action.

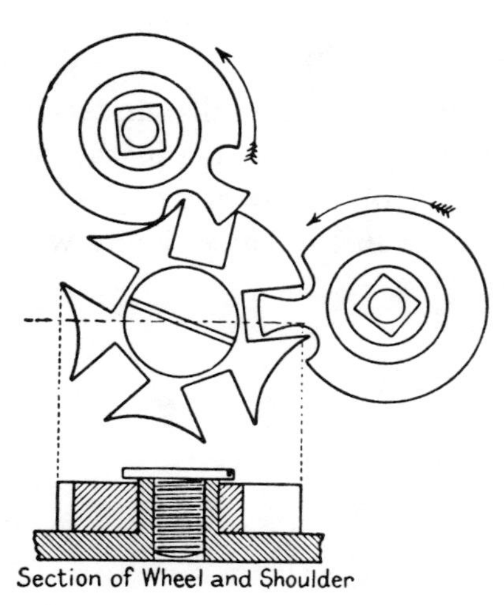

FIG. 82.—A high-grade form of watch stop-work. The star wheel rides upon a brass pipe in the plate instead of being kept in place with a shoulder screw.

Hooking-in. The formulæ given in Chapter VII for determining the dimensions of clock mainsprings and the number of turns of winding are equally applicable to watch mainsprings. Experience has taught that the barrel arbor should not, however, be less in diameter than one-third that of the barrel.

The method of hooking-in requires a certain amount of careful attention. With the going barrel, the outer coils can be made to offer considerably less resistance when fully wound if the end of the spring is softened and bent outwards to form a small hook. A short length of the spring is then inserted between the two hooks, the one on the end of the spring and the other in the barrel edge. This is the best form of hooking-in to ensure central development of the coils. In another form which avoids any wrenching a piece riveted on the end of the spring has two pivots, above and below, the bearings for which are holes drilled in the cover and barrel bottom. Both these forms are known as "yielding attachments," in contrast to the "rigid attachment" of direct hooking used in fusee watches. In these, a square hook carefully undercut and riveted to the spring enters a prepared hole in the edge of the barrel, and the outer coil of the spring is then held fast against the barrel, no space in the barrel being wasted.

CHAPTER XVI

KEYLESS MECHANISMS

It appears to be extremely difficult to credit anyone in particular as having originated the idea of winding a watch without using a key. In this country, however, the practical method of applying the principle differed in every possible way from that developed on the Continent. There was not the least similarity between them. To-day the principles involved in keyless-mechanism construction virtually fall into five different groups. Each, however, may be and is applied in a variety of ways by different manufacturers, suiting the grade and finish of the watch whereof it forms a part. The groups are as follows :—

1. English rocking bar.
2. English fusee-keyless.
3. Continental top-stem wheel and sliding pinion, with push-piece or slide hand-setting.
4. Continental top-stem wheel and sliding pinion, with pull-out-piece hand-setting.
5. American and Continental negative-set.

Rocking-bar Mechanisms. Most English movements are provided with the form of keyless mechanism shown in Fig. 83. The "rocking bar" is a steel frame upon which three steel wheels are mounted to gear with one another. The centre or crown wheel, *C*, about which the rocking bar is free to oscillate, is larger than the two side wheels, S_1, S_2. The crown wheel is immediately below the pendant, and a bevel gearing is formed between it and a vertical driving pinion, *D*, the stem of which passes through the pendant and is squared to receive a winding button. The top of the button is "cut" or recessed to admit a small nut, which is screwed to the extreme end of the stem, leaving just sufficient freedom for the smooth motion of the driving pinion. The backs of the crown-wheel teeth are slightly

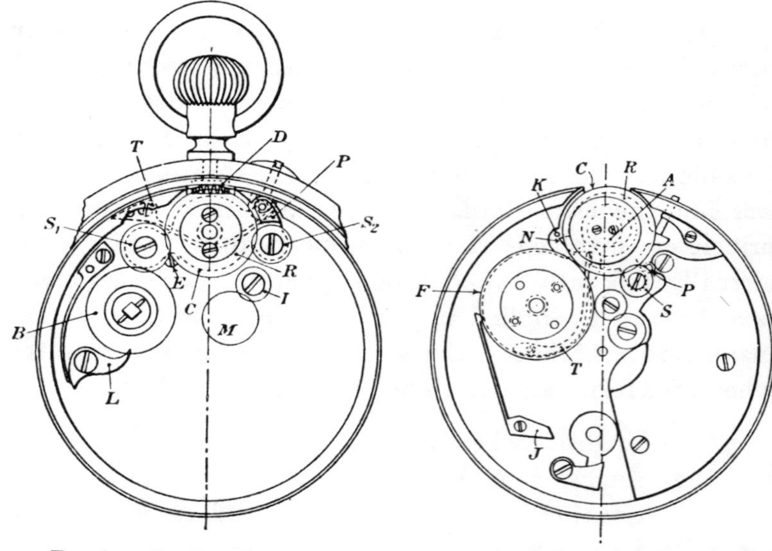

FIG. 83.—English Rocking-bar keyless mechanism.

FIG. 84.—An efficient form of English Fusee-keyless mechanism.

bevelled to ensure good engagement with the driving pinion. The movement of the rocking bar is restricted by the motion of the projecting piece, *P*, passing through a slot in the frame, but it is kept normally at the winding position by the return spring, *T*. That is to say, looking at the mechanism bottom-plate uppermost, as in the figure, the side wheel, S_1, engages with a steel barrel ratchet, *B*, which is squared on a projection of the barrel arbor, the depth being controlled by the screw, *E*, and the winding maintained by the click, *L*. The design of this click should be noticed, because it acts upon ordinary wheel teeth and not on the undercut teeth appearing in most forms of click and ratchet works. Correct planting of this click is also a matter of extreme importance, otherwise it may fail. A sufficient amount of recoil must be allowed and the nose requires careful adjustment. When the motion of the winding button is reversed, the back action is produced by the movable rocking bar yielding to the stationary barrel ratchet, *B*, the tension of the spring, *T*, causing the side wheel to slide over the barrel ratchet one tooth at a time. The push-piece in the band of the case presses on the projection, *P*, when the hand-setting is required. This causes the side wheel, S_1, to become disconnected from the barrel ratchet, *B*, and the side wheel, S_2, to engage with the idle wheel, *I*, which is permanently in gearing with the minute wheel, *M*.

Fusee-keyless Mechanisms. As the name implies, the object of the fusee-keyless mechanism is to incorporate the advantages of the fusee without involving the necessity of using a key for the purpose of winding. This combination involves a delicacy of mechanism which has confined its application to high-grade watches. Many ingenious devices are met with, some more complicated than others, besides being unfortunately often unreliable in their action. The underlying difficulty is that the fusee arbor revolves in unwinding so that, if it were combined with an ordinary keyless mechanism the winding button, would revolve too. This, of course, could not be allowed to happen in the pocket, or the watch would stop. The method shown in Fig. 84 is, however, not only extremely simple in principle, but perfectly reliable in practice. But for the fact that very careful workmanship is essential, its construction also does not present serious difficulty.

The rocking bar, *R*, carries two wheels only and is free to move in a horizontal plane as well as about its own centre. The crown wheel, *C*, is recessed and is only partly covered by a shoulder cap, *A*, which is fastened with two screws to the bottom plate, whilst the return spring, *T*, throws the whole rocking bar over to the right against the cap. The effect thus produced is an eccentric movement of the wheel when pressure is applied through the driving pinion in the pendant.

The fusee arbor projects through the frame and is tapped to carry a steel winding wheel, *F*, corresponding to the barrel ratchet in the usual mechanism. Into this wheel the crown wheel engages only when driven by operating the winding button; normally it is out of action because of the return spring, *T*. The actual depth of engagement of the two wheels, *F* and *C*, is controlled by the nose, *N*, which projects from the rocking bar and acts against the banking pin, *K*. This brings the rocking bar to rest when the nose has advanced until the pin reaches the butting corner.

It will be seen that the slow motion of the fusee in unwinding would present occasions when the crown wheel could not engage smoothly into the fusee wheel and the tips of the teeth would butt. This difficulty is obviated by a supplementary device on the winding wheel, *F*. Instead of this wheel being screwed directly to the fusee arbor, it is recessed on the upper side and screwed to a cap, the centre of which is colleted and tapped to take the fusee arbor. In screwing the wheel to the cap just enough shake is allowed to give a slight movement to the wheel. When in position, this wheel is held tooth by tooth, by a jumper spring, *J*, and thus it is always able to offer immediate admission to the crown wheel without any danger of injury to the teeth. The small side wheel, *S*, engages in the usual way with the motion work when the rocking bar is depressed by operating the push-piece.

The back action, however, is in this case determined by the pin, *P*, which is the polished end of a screw projecting from the underside of the plate. When the motion of the winding button is reversed, the teeth of the side wheel impinge against the pin, *P*, under the tension of the rocking-bar return spring, *T*, and click over, tooth by tooth, in passing.

Top-stem wheel and Sliding pinion Mechanisms with push-piece Hand-setting. As in the case of the English rocking bar, it is difficult to ascribe definitely the origin of the Continental keyless mechanism to any one person. Fig. 85 illustrates one of the earlier forms which became generally adopted in good watches and the principle has been maintained ever since with but minor modifications. Fig. 85A is a view of the front plate, whilst Fig. 85B shows the frame reversed. The shaft or stem, *A*, passes through the pendant of the case and carries the button, whilst at the other end it is pivoted to work into a hole in a brass block, either part of or screwed to the bottom plate. The stem is of round section of varying diameters as far as the top-stem wheel, *B*, which is often known as the "Bréguet wheel," sometimes the "crown" and also the "transmission." Below this, the arbor is squared to carry the sliding-pinion, *C*, also termed the "castle-wheel," and is sufficiently long to allow this pinion a certain amount of up-and-down movement. The top-stem wheel has two sets of teeth, one flat and the other, on the underside, ratchet-shaped, to form a right-angle gearing. The sliding pinion has also two sets of teeth, ratchet-shaped on the top to gear with the top-stem wheel, and ordinary or bevelled teeth underneath to engage with a small pivoted transmission wheel, *T*. The figure shows two transmission

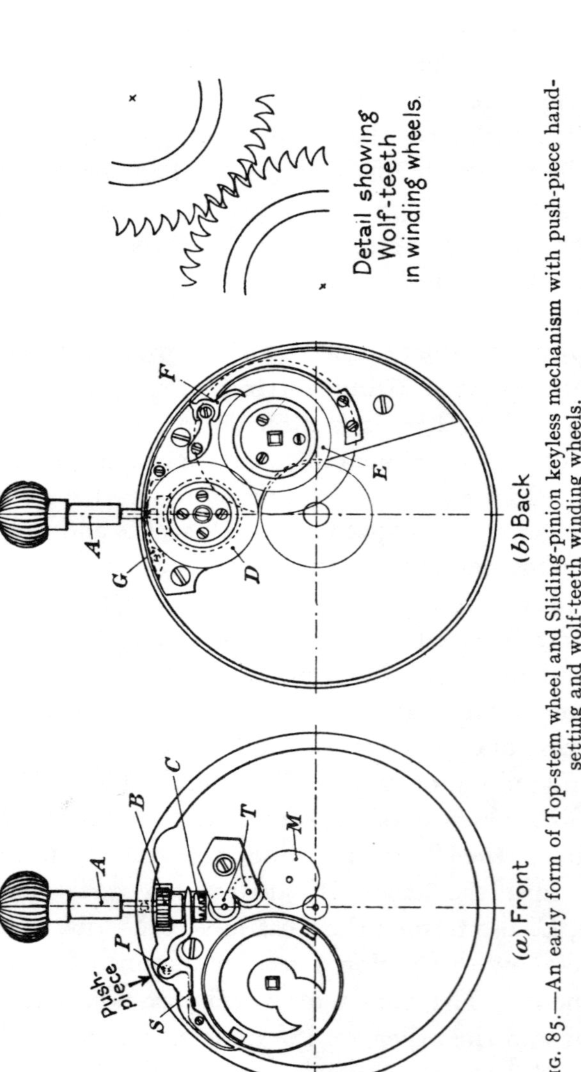

FIG. 85.—An early form of Top-stem wheel and Sliding-pinion keyless mechanism with push-piece hand-setting and wolf-teeth winding wheels.

wheels permanently in engagement with the minute wheel, *M*. In some instances only one is used. Round the body of the sliding pinion a channel or groove is provided for the purpose of admitting the end of the set-hand lever, *S*, which is steady-pinned and screwed to the plate at one end and so formed that it acts as a spring to keep the sliding pinion normally taut against the top-stem wheel. If, however, a push-piece in the band of the case is depressed the set-hand lever is also depressed where indicated by the dotted circle, *P*, carrying with it the sliding-pinion on the squared arbor of the stem. This enables the sliding-pinion to engage with the transmission wheel so that the hands may be set. Releasing the push-piece immediately causes the sliding-pinion to return to its normal position in gear with the top-stem wheel. Fig. 85B shows the movement reversed with the barrel bar screwed to the bottom plate and two winding wheels, *D* and *E*, gearing into each other on the barrel bar. *E* is the barrel ratchet squared and screwed by three screws to the barrel arbor. As the mainspring is wound the clickwork, *F*, prevents the wheel running back and directs the motion into the train. *D* is a double-cut crown wheel, the teeth on the face engaging with the ratchet, whilst a series of teeth project below the bar to gear into the top-stem wheel teeth, thereby completing the communication between the winding button and the mainspring barrel. The crown wheel, *D*, is covered by a cap which when screwed to the bar allows the wheel correct freedom. The stem is fastened in place by a stem-bridge, *G*, springy at one end, so that when screwed home it enters a channel cut in the stem, and though the stem is allowed sufficient freedom it is prevented from being pulled out. A screw, or sometimes one with a long pivot, may be substituted for the stem-bridge.

It is quite customary to provide watches, especially the

better grades, with winding wheels cut with wolf teeth as shown in detail at the side of the figure. This renders the teeth extremely durable, the acting faces being involute in form though designed to operate in one direction only.

Top-stem wheel and Sliding pinion with pull-out-piece Hand-setting. Very many modern watches are not provided with a push-piece in the band of the case to press when the hands require to be set. In place of this system, hand-setting is effected by pulling out the winding button

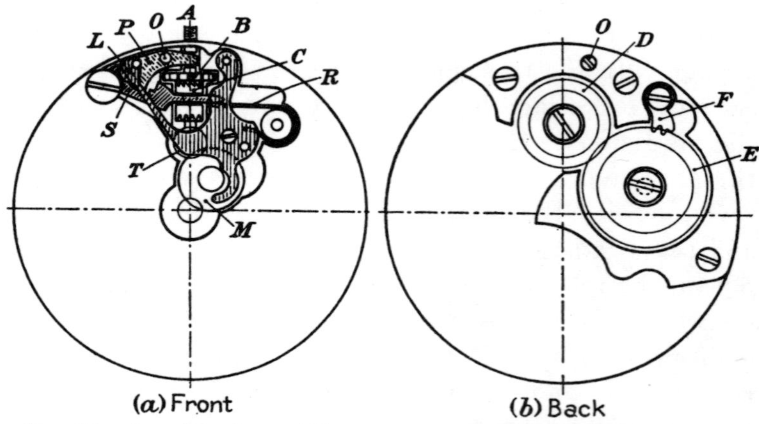

FIG. 86.—A modern form of Top-stem wheel and Sliding pinion keyless mechanism with pull-out-piece hand-setting. (See also Fig. 87.)

a short distance, and the mechanical medium for changing the position of the wheels for doing this is called the "pull-out-piece." The application of this, of course, varies with the designs of different makers, but Figs. 86 and 87 may be regarded as typical and well-designed examples. In Fig. 86 the stem, *A*, and the top-stem wheel and sliding-pinion, *B* and *C*, are similar to the mechanisms shown in Fig. 85, but the set-hand lever, *S*, is not partly a spring in this case and it is simply held at one end by a shoulder screw about which it is free to move. It is formed, however, in such a way that part of its upper edge can be depressed by the movement of the pull-out-piece, *P*. This is a cranked lever with a tapped hole at *O*, to take a carefully fitted shoulder screw projecting through the barrel bar from the back. The body of this screw is surrounded by a shoulder so that it is prevented from moving up and down between the frame, though free to tighten or release the pull-out-piece. One end of the lever is bent at right angles or blocked to enter with sufficient freedom a channel turned in the winding stem, whilst the other end is tapered off at such a length as to remain always in proximity to the upper edge of the set-hand lever. This edge of the set-hand lever is curved and contains a small notch. The front of the set-hand lever is narrowed down to intercept the channel in the sliding pinion, as before, and a return spring, *R*, acting on the extremity keeps the two wheels in safe contact. At the same time, the pull-out-piece is kept in position by a pin, which drops into a hollow provided in the end of the locking or bolt spring, *L*. As the pull-out-piece is drawn up by the winding stem when the hands require to be set, the tapered end, acting about the centre, *O*, depresses the edge of the set-hand lever until it enters the notch, and the pull-out-piece having performed its maximum amount of movement is locked. The angle through which the set-hand lever passes from the winding position to the notch is proportionate to the angle of motion required to lower the sliding pinion to engage correctly with the transmission wheel, *T*. The minute wheel, *M*, is in permanent gearing with the transmission wheel. The crown wheel, *D*, and barrel ratchet, *E*, in Fig. 86B connect the stem with the barrel and mainspring, the clickwork, *F*, taking the strain.

Fig. 87 shows the same principle applied in a rather different way. The locking-spring, *L*, has a double notch which secures the pull-out-piece, *P*, at the winding and hand-setting positions respectively. The tail of the pull-out-piece slides up an inclined plane end to the set-hand lever, *S*, in order to lower the sliding pinion, *C*, into the transmission wheel, *T*. In this design, the set-hand lever approaches the sliding pinion from the opposite side to that shown in Fig. 86.

In both illustrations, a form of recoiling click, *F*, is shown which possesses the advantage of preventing the coil of the mainspring from binding through being overwound. These clicks have two or three teeth and, in process of winding,

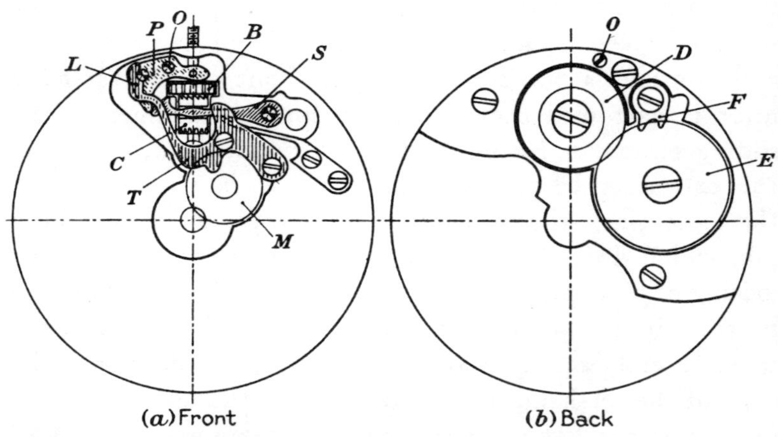

FIG. 87.—Another example of Top-stem wheel and Sliding pinion keyless mechanism with pull-out-piece hand-setting. A different application of the principle to that shown in Fig. 86.

the fact that both or all disengage from the ratchet provides a sufficient amount of backlash in re-engaging to ease slightly the tension of the mainspring.

American Negative-set Mechanism. Although now extensively applied to Continental watches, this method of hand-setting is very intimately associated with American productions. The system sprang out of the custom of completely separating the manufacture of the movements and cases. Both being standardised to suit one another, cases could be made if necessary in countries where movements could not, an economical procedure which would tend to lessen the cost of the finished article. The case has a tapped pendant as shown in Fig. 88, screwed into which is a steel sleeve, *S*, with four claws or tongues. The centre of the sleeve is hollow to allow the stem, *R*, to pass through it. One end of the stem is tapped to carry the button and the other end squared to project towards the movement within the band of the case. Where the stem intersects the edges of the sleeve claws, however, there is a channel, and below this is a V-shaped nick into which the claws spring. When the winding button is pulled out or pushed in the stem is limited in its movement to the length of the V-nick.

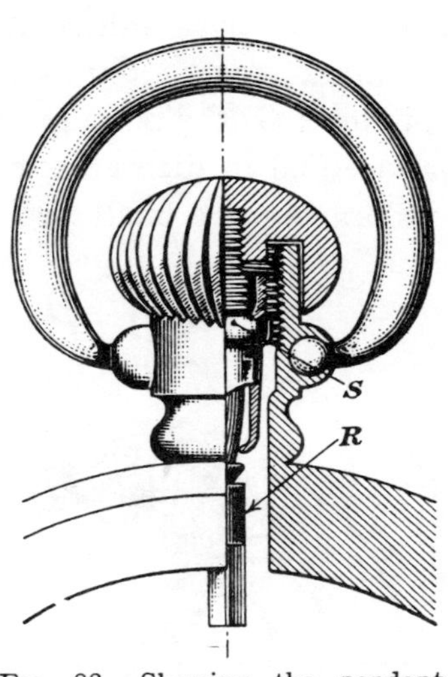

FIG. 88.—Showing the pendant and sleeve construction in the Negative-set system of keyless mechanism.

This figure also illustrates an interesting safety method of securing the bow to the pendant, which is adopted by the Dennison Watch Case Co., who have so largely developed in this country the manufacture of cases to suit standardised negative-set movements.

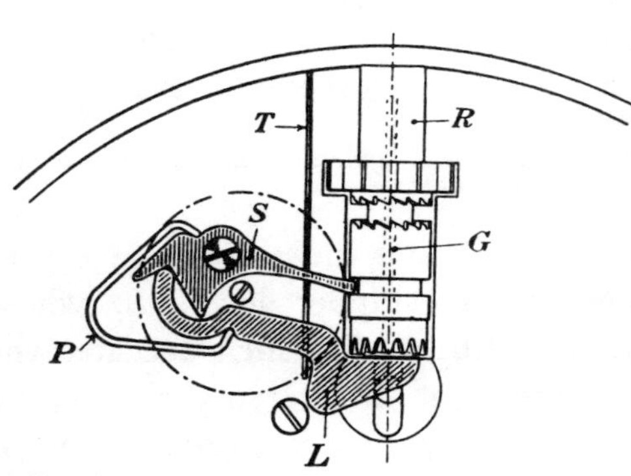

FIG. 89.—Negative-set keyless mechanism of the Waltham Watch Company.

Fig. 89 shows a most ingenious keyless mechanism,

introduced by the Waltham Watch Company. The stem, *R*, does not project beyond the frame and has a squared centre to receive with sufficient freedom the squared end of the pendant stem. Through the centre of the stem passes a plunger, *G*, the bottom of which is of screw-head formation and equal in diameter to the pivot of the stem, which has its bearing in a small steel collet screwed to the frame. The

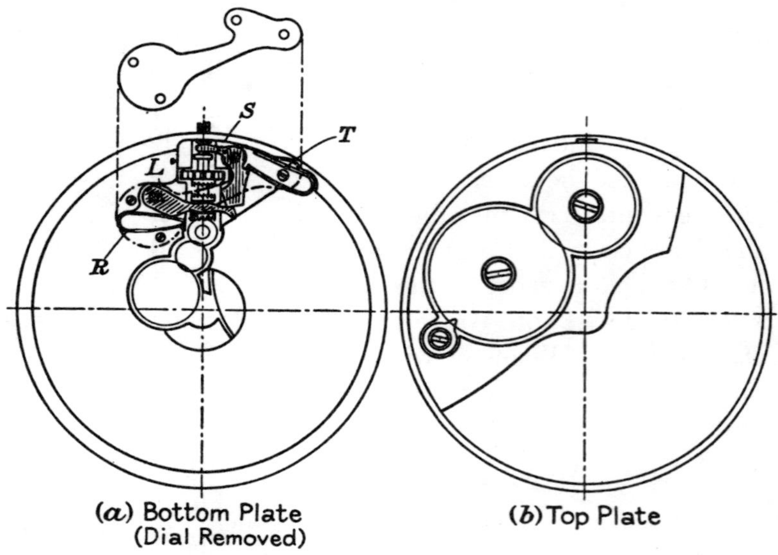

FIG. 90.—An efficient form of Negative-set keyless mechanism adopted in Continental watches.

head of the plunger impinges on the shipper lever, *L*. The curved formation of the shipper lever, *L*, where contact is made with the shipper, *S*, which actuates the sliding pinion, has the effect of maintaining the sliding pinion normally at the hand-setting position, *S* and *L* being clamped together by means of the shipper spring, *P*. When the plunger is depressed by the stem in the pendant, the shipper lever, acting against a banking screw, throws up the shipper and raises the sliding pinion into gear with the top-stem wheel for winding. *T* is known as the shipper bar and has a projection at right angles both top and bottom for use only to draw up the shipper if it is desired to run the watch out of the case.

Continental Negative-set Mechanism. Many varieties of negative setting are now adopted in Continental watches. The form shown in Fig. 90 may be regarded as a design which is both reliable in its action and simple in construction. The setting mechanism consists merely of two crank levers and two return springs covered, when in position, by a suitably-shaped cap screwed to the plate. These levers and springs are respectively known as the set-hand-piece, *S*, actuated by the thrust spring, *T*, and the set-hand lever, *L*, actuated by the return spring, *R*. A definite locking is provided in the shape of the contact surfaces between the set-hand-piece and the set-hand lever. The pendant work in the cases to which these watches are fitted is similar to that shown in Fig. 88.

CHAPTER XVII

CHRONOGRAPHS AND STOP WATCHES

THE very apparent advantage of being able to measure small intervals of time under certain circumstances led to the introduction of centre-seconds watches. In these watches the fourth pinion is planted in the centre and carries, through the dial, a long pivot on which is mounted the seconds hand. This extends to the minute circle on the dial which is subdivided into fifths and serves also to indicate seconds. A slide in the band of the case affords an opportunity of stopping the mechanism at any instant by thrusting a fine wire, often a piece of balance-spring, into the path of the balance or against the roller. The unpleasant feature of thus entirely stopping the watch was, however, removed on the appearance of an improved form of independent centre-seconds movement, where two separate trains, one for going and one for seconds, were employed.

Centre-seconds stop watches must, nevertheless, be regarded as having been now superseded by the fly-back chronograph for use whenever short intervals of time require to be noted. The name appears to some extent to seem a misnomer, because anything described as a "graph" would lead one to expect the instrument to make its record on paper or a chart. The fly-back chronograph, however, is not possessed of such powers and merely enables the operator to start and stop the seconds at will, whilst the train keeps going, and lastly to restore the seconds hand to zero ready to make a fresh start. It would really be more correctly described as a chronoscope. In point of fact it should be mentioned that early examples of pocket chronographs were constructed as ink recorders.

Minute-recording Chronographs. There is some variety in the design of these watches, but Fig. 91 may be accepted as a typical example of a Swiss minute-recording

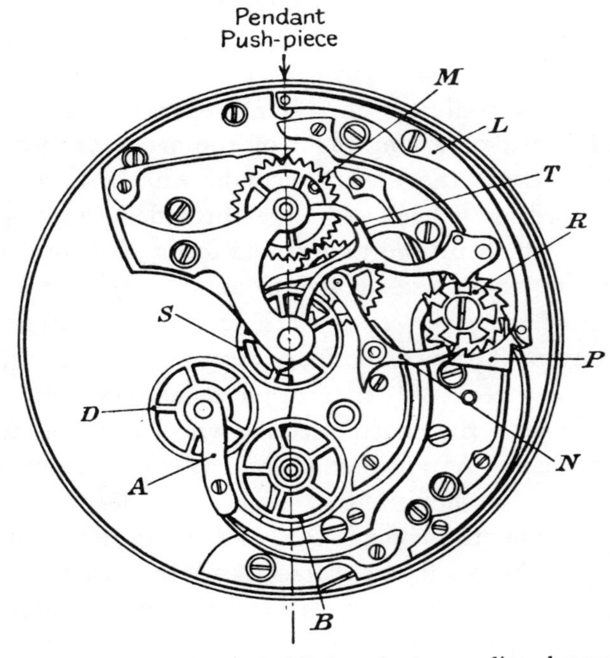

FIG. 91.—Mechanism of a typical Swiss minute-recording chronograph.

chronograph. The mechanism is started by pressing a push-piece which passes through the winding button, thus deflecting the start lever, *L*. Being free to move about a shoulder screw, the other extremity of this start lever carries a pivoted pawl, *P*, and pulls over a steel ratchet, *R*, upon the surface of which and at every third tooth rise six turrets, each the width of one ratchet tooth. A wheel, *B*, cut with fine involute teeth, riding on the fourth pivot which comes through the plate, gears permanently with the intermediate wheel, *D*, of the same size and number. This wheel is pivoted at the extremity of the lever, *A*, which is actuated by the movement of the turrets on the steel ratchet, *R*. When the intermediate wheel, *D*, is brought thus into engagement with the centre-seconds wheel, *S*, also cut with fine teeth, the recording commences, and continues until the two are separated by the succeeding operation of the pendant push-piece. The minute-recording hand which indicates on a small separate circle on the dial, is mounted on a long pivot of the arbor carrying the wheel, *M*. Both the centre-seconds wheel, *S*, and the minute-recording wheel, *M*, are provided with cams, pressure on which by the two-armed lever, *T*, causes them to return to their zero position. Hence with the first depression of the lever, *L*, a tooth of the steel ratchet, *R*, is gathered up, the lever, *A*, drops between two turrets, the intermediate wheel, *D*, at its other extremity passes into engagement with the centre-seconds wheel and the recording commences. At another part of the steel ratchet, *R*, the lever, *T*, has mounted a turret and cleared its two extremities from the cams. One end of the lever, *N*, has also dropped between two turrets and thrown the intermediate-recording wheel, pivoted at its other extremity and in permanent gear with the minute-recording wheel, *M*, slightly nearer the centre-seconds wheel, *S*. It is thus placed in the path of a finger which is attached to the centre arbor and advances one tooth of the intermediate wheel with every complete revolution to which the minute-recording hand responds accordingly. The second depression of the lever, *L*, causes *T* to remain in the same position, but allows *A* to mount the next turret and disengage the intermediate wheel, *D*, from the centre-seconds wheel, *S*, and the recording ceases. The third operation pulls the turret clear of the lever *T*, and as it falls into the space the cams on *S* and *M* return the respective hands to zero, the lever *N* having first removed the intermediate-recording wheel from the path of the centre-seconds finger. The lever *A* is left riding on the turret. All the pieces act against return springs. The centre-seconds wheel has a

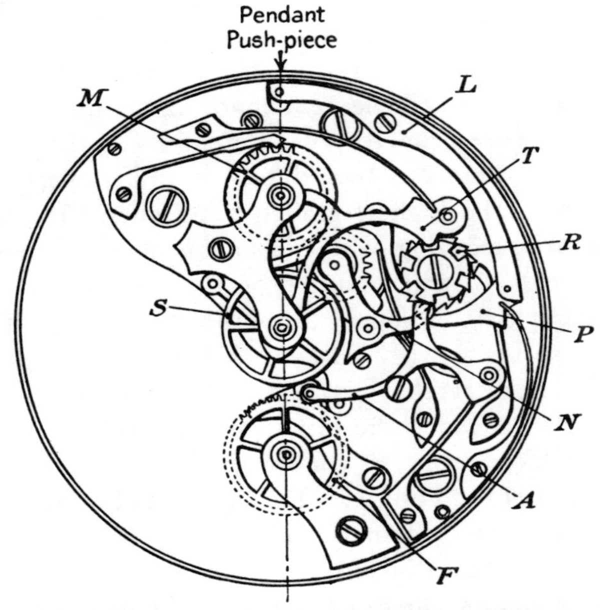

FIG. 92.—A form of Swiss minute-recording chronograph mechanism in which a double-headed pinion, pivoted at the extremity of the lever, *A*, is used in place of the two fine-cut wheels shown in Fig. 91.

slender spring brake to keep it friction-tight between the frames, and the minute-recording wheel, *M*, is held in its place by a V-jumper, or "recorder detent."

Another design of chronograph movement frequently met with is shown in Fig. 92. In this case, two of the fine-cut wheels are dispensed with and, in their place, a double-headed pinion is pivoted between the bottom plate and the extremity of the lever, *A*. One head of the pinion is in permanent engagement with the fourth wheel, *F*, but the other is finely cut to engage directly with the centre-seconds wheel, *S*, when necessary. Though to some extent differently formed, the other parts of the mechanism closely resemble that previously described and the sequence of operations is also similar.

The minute-recording mechanism in both these examples is known as "creeping recording," because the centre-seconds finger has to advance the intermediate-recording wheel gradually until the next tooth of the minute-recording wheel passes the recorder-detent. Another method, called "instantaneous recording," is preferable, but only met with in the highest grades of watches. In this case, the minute-recording wheel is an ordinary ratchet of 30 or 60 teeth. A snail mounted on the centre-seconds arbor in rotating lifts a lever, the other extremity of which carries a spring pawl. This action causes the pawl to ride over the next ratchet tooth with sufficient clearance to enable it to advance the minute-recording wheel immediately the lever drops off the undercut step of the snail on completion of the minute. This method of minute-recording will be observed in the illustration of a split-seconds chronograph shown in Fig. 93.

Split-seconds Chronographs. The design of chronographs became more complicated on the introduction of the split-seconds. Not only can the centre-seconds mechanism be started, stopped and returned at will, but by means of an auxiliary push-piece an additional centre-seconds hand can, without impeding the progress of the other, be stopped at any instant and, after making the necessary observation, be advanced rapidly to catch up the first hand with which it proceeds as before. So long as the two hands are running together, however, they respond as one to the operation of the pendant push-piece.

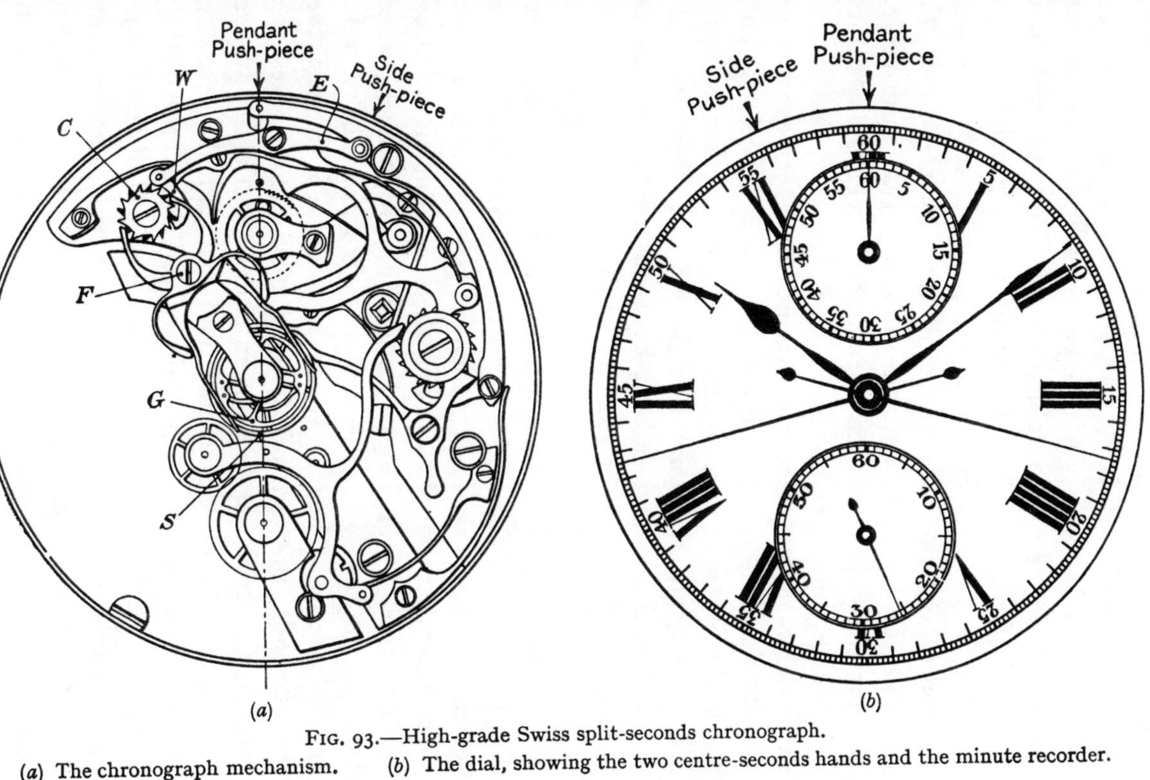

FIG. 93.—High-grade Swiss split-seconds chronograph.
(a) The chronograph mechanism. (b) The dial, showing the two centre-seconds hands and the minute recorder.

An illustration of a high-grade movement is given in Fig. 93, where it will be seen that an extra steel ratchet, *C*, with turrets is used. In this particular case, the turrets are simply formed by bevelling off the alternate teeth of a very thick ratchet. The bevelling only extends through half the thickness, the teeth being intact in the lower portion where they are operated upon by the pawl, *W*, through the lever, *E*, and the auxiliary side push-piece. The ordinary part of the chronograph mechanism (Fig. 93A), chiefly on the right-hand side, is similar to that shown in Fig. 91. The extra mechanism, however, consists of an additional heart-cam, the split-seconds wheel, *G*, and the brake system connecting these with the steel ratchet, *C*. The heart-cam is mounted tight on the centre-seconds arbor, in this instance hollow, between the centre-seconds wheel, *S*, and the split-seconds wheel, *G*. Pressing constantly against this cam is the edge of a jewel hole pivoted at the end of a short lever, which is attached at the other end by means of a shoulder screw to the wheel itself. The jewelled lever always finds the lowest part of the heart-cam carrying with it the split-seconds wheel, *G*, which is mounted to run freely. A long arbor from this wheel passes through the hollow centre-seconds arbor and dial to take the split-seconds hand. When it is desired to stop the split-seconds, pressure of the side push-piece causes the ratchet, *C*, which has an even number of teeth, to advance one tooth. At the same time, two diametrically opposite turret teeth part from the tails of two long cross-acting levers, *F*, the fronts of which grip both sides of the finely-serrated split-seconds wheel, *G*. Light springs control the grip of the arms and, whilst the wheel and the jewelled lever underneath remain stationary, the latter keeps in contact with the edge of the heart-cam. On the next operation of the side push-piece the wheel is released by the cross-acting levers, and the jewelled lever moves instantly to the deepest part of the heart-cam accompanied by the wheel with the duplicate seconds hand. Fig. 93B shows the dial with the two centre-seconds hands and minute recorder.

Timers. For rendering the specific service of timing short intervals "timers" or "recorders" have proved

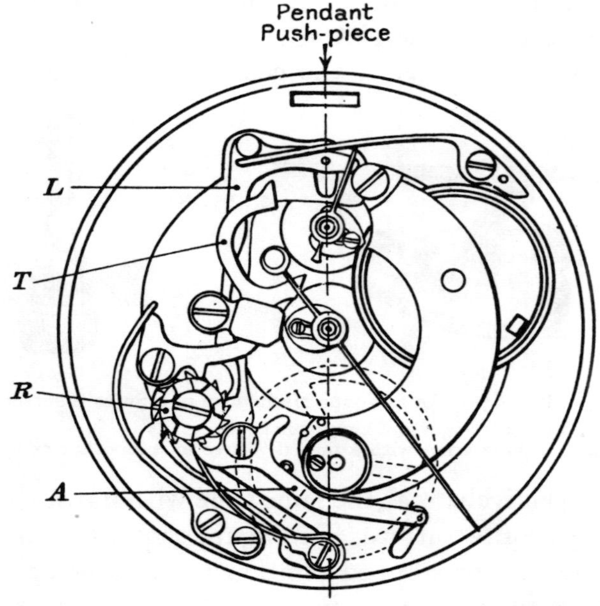

Fig. 94.—Start, stop, and fly-back mechanism of a Swiss tenths-second timer.

themselves a very useful supplement to the orthodox chronograph. In the timer, the ordinary 30-hour going with hour-and-minute motion work is entirely abandoned and the mechanism is stopped by a pin or length of balance-spring impinging against the rim of the balance as in the old centre-seconds watches. These instruments are generally inexpensively constructed, though efficient enough for timing short periods in fifths or tenths of a second to meet usual requirements. Some are even constructed to show hundredths of a second. Fig. 94 illustrates the start, stop and fly-back mechanism arranged under the dial of a tenths-second timer. On depressing the button, the lever, *L*, propels the turreted ratchet, *R*, and the lever, *A*, is deflected by its tail being admitted into the space between two turrets. This action releases the hold, on the rim of the balance, of a fine brass pin inserted in the end of lever *A*, and passed through a hole in the plate. The zero lever, *T*, which has two arms acting against the centre and minute-recording

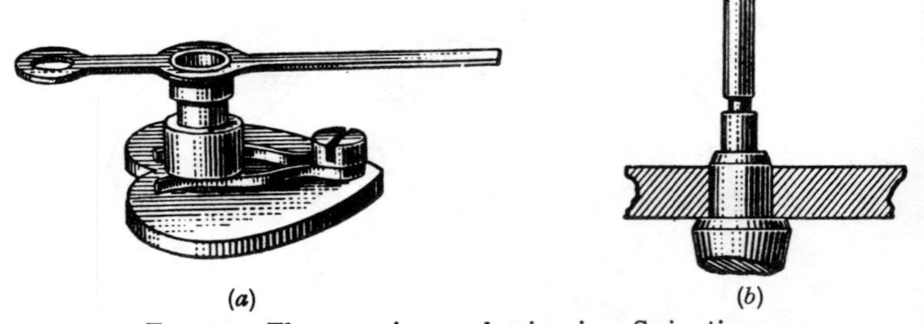

(*a*) (*b*)

FIG. 95.—The cam-pipe mechanism in a Swiss timer.
(*a*) The heart-cam and keyed pipe.
(*b*) The long arbor with a turned groove to receive the cam-pipe.

cams respectively, is at the same time withdrawn by mounting an opposite turret. The figure shows the mechanism in the second or stopped position prior to being returned to zero at the third operation.

The train is arranged so that the barrel drives directly into a pinion with a long arbor carrying a heart-cam and the minute-recording hand. When the mechanism is running, this hand consequently proceeds continuously instead of being advanced only once a minute and makes one complete rotation in 15 minutes. The wheel mounted on the minute-recording pinion gears with the centre-seconds pinion through an intermediate wheel and pinion and the centre-seconds wheel runs into the escape pinion. To record tenths of a second the balance is sprung accordingly and the centre-seconds hand makes a complete revolution of the dial in 30 seconds. The arrangement of the fifths-second recorder is exactly similar, except for the fact that it is sprung as for the usual 18,000 train and the centre-seconds hand goes round the dial only once in a minute.

The heart-cams (Fig. 95A) in these mechanisms carry short pipes to enable them to be mounted on the respective arbors (Fig. 95B), but they are rendered friction-tight by means of a double-acting spring (Fig. 96). The spring is

Fig. 96.—A sectional view of the cam-pipe and arbor shown in Fig. 95.

screwed to the cam near the pointed end with two arms projecting towards the pipe. There is a groove turned in the arbor and a keyway cut correspondingly in the pipe, so that one arm of the spring enters the keyway and groove pressing against the arbor whilst the other arm presses on the outside of the pipe itself. The tension of the spring is thus sufficient to maintain the motion of the heart-cam and pipe with its hand, when the train is running, but not enough to withstand the force of the zero lever against the heart-cam at the returning operation.

Split-seconds timers are also made on similar lines and dials are painted in a variety of ways to suit different requirements and purposes.

CHAPTER XVIII

REPEATING MECHANISMS

Early Designs. Striking watches, more generally known as clock watches, were in use in this country as far back as the sixteenth century, and the principle generally adopted for this purpose was similar to the locking-plate striking mechanism of a clock described in Chapter VIII. With an independent train and a lifting-piece actuated once an hour by a nose-piece fastened to the cannon pinion, the hammer blows were struck on a bell which covered the back and side of the movement. Plate IX illustrates, in three positions, a clock-watch movement of about 1685, by Samuel Marchant. The first is a front view with the dial removed, leaving the lifting-piece plainly visible. The second is a side view showing prominently the striking mainspring barrel, as well as the warning and stop levers, all of which are richly pierced. The large bell which covers the whole movement is also shown. The third position presents a view of the back, where the locking plate can be seen. The locking-plate wheel, which is driven by an end pinion just as in a clock, is beneath the locking-plate.

Towards the end of the seventeenth century this method of hourly striking, which possessed many disadvantages, was destined to be superseded by an entirely new system of repetition striking at will.

In principle this form of striking for watches closely resembles the rack mechanism for clocks and it seems

PLATE IX.—A "clock-watch" movement of about the year 1685, by Samuel Marchant, London. This early specimen of a striking watch is constructed on the principle of the locking-plate striking mechanism for clocks. The locking-plate is visible in the back view.

[*To face page 220.*

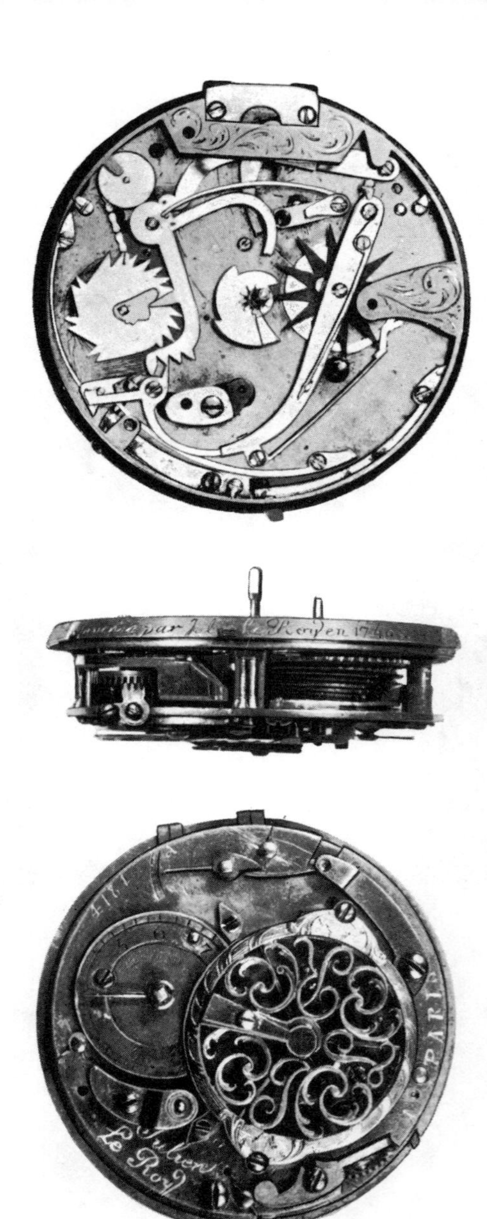

PLATE X.—The mechanism of an early repeater by Julien Le Roy, showing (in the front view) the "all-or-nothing" piece which appears to have been originally invented by him. The maker's initials can be seen pierced on the balance-clock.

[*To face page 221.*

probable that the two originated about the same time. The idea appears to have been prominent in the minds of several makers of the day, though Daniel Quare (1649–1724), celebrated for his fine workmanship, is credited with the invention in the year 1676.

In older models of repeating mechanisms the pendant was pressed in to wind the repeater mainspring as desired, the two being connected by a stout fusee chain. The hammer blows were either given on a bell or dumb block in the case. As their popularity increased they were taken up by Continental makers and two distinct improvements are attributed to the efforts of a renowned Paris horologist, Julien Le Roy (1686–1759). Hitherto, if the pendant was not pressed in to the full extent, only a portion of the rack passed the hammer pallets and the striking would not be correct. Le Roy seems to have been the first maker to introduce a separate device, known as the "all-or-nothing" piece, which prevented any striking taking place unless the pendant was pushed in to the fullest extent, thereby ensuring that the rack was correctly set.

Another feature alleged to have been introduced by Le Roy, was striking on gongs composed of circular wire springs, giving a different tone for the hours and quarters, in place of the bell or dumb block.

Three views of a specimen of Le Roy's productions are given in Plate X. The first is a front view showing the repeating work, to which further reference will be made later. The second shows the one hammer which strikes a dumb block in the case, and the third displays the back with the maker's initials pierced in the bridge as well as a novel device for regulating the speed of the repeater train.

An interesting feature about the movement is the inscription on the brass edge to the front plate, partly visible in the second view. This reads *Inventé par Jul. Le Roy en*

1740 *et Le Rouage* 1755, which would appear to signify that he designed the train after he had solved the problem of the repeating mechanism.

In modern repeaters, the pendant push has given place to a slide or, in cheaper grades, a push-piece at the side in the band of the case; and instead of a length of chain connecting this to the barrel a circular winding rack is in direct engagement with the barrel arbor or an intermediate wheel.

There are four varieties of repeaters :—

Quarter Repeaters, giving hours and quarters.

Minute Repeaters, giving hours, quarters and minutes.

Half-quarter Repeaters, giving hours, quarters and half-quarters.

Five-minute Repeaters, giving hours and five-minute intervals between the hours.

The last two are not frequently met with nowadays.

The repeater train terminates in a speed-regulating device which may take the form of an eccentric bush whereby the motion is retarded or freed by the adjustment of the depth of the last wheel and pinion. Otherwise, the train may be controlled by a recoil escapement with an adjustable banking, or it may be effected by means of a centrifugal governor.

Quarter Repeaters. Fig. 97 illustrates the mechanism of a modern high-grade Swiss open-face quarter-repeater and the sequence of operations may be enumerated as follows. In Fig. 97A the slide, *S*, pushes round the winding rack, *W*, carrying with it a pivoted arm, *A*, the end of which butts against the hour snail situated under the star wheel, *V*. This arm is provided with a trigger or release piece, *T*, the edge of which follows round the inner surface of the all-or-nothing piece, *N*. As soon as the arm, *A*,

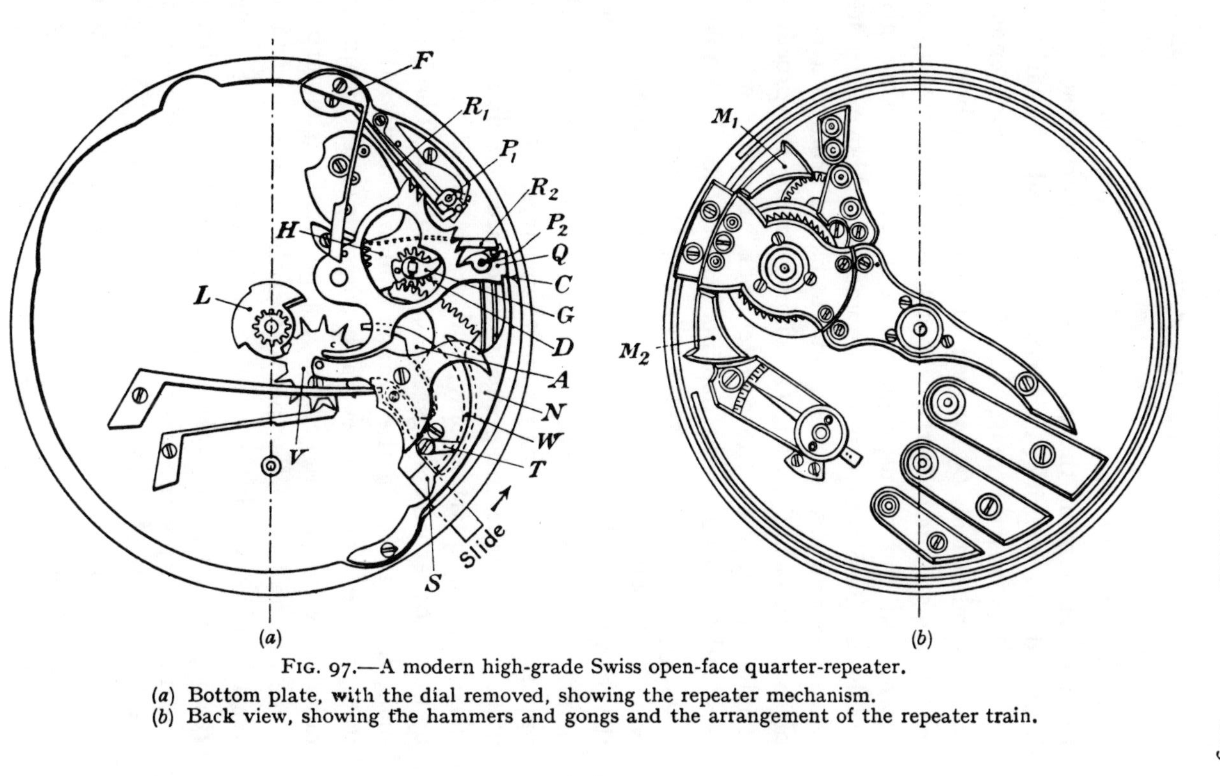

FIG. 97.—A modern high-grade Swiss open-face quarter-repeater.

(a) Bottom plate, with the dial removed, showing the repeater mechanism.
(b) Back view, showing the hammers and gongs and the arrangement of the repeater train.

reaches the hour snail the trigger, *T*, is tightened against a stop screw and deflects the all-or-nothing piece. At the same instant, the end of the all-or-nothing piece having been released from the notch, *C*, the quarter rack, *Q*, is liberated and a spring, *F*, carries it over till the tail butts against the quarter snail, *L*. The winding rack engages with a steel wheel, *D*, squared on the barrel arbor. This wheel has an uncut portion to act as a stop both up and down, and the mainspring, when in full use, thus gives nearly a complete turn in unwinding. Above the steel wheel, *D*, and also squared on the barrel arbor is a hook-piece or gathering pallet, *G*, which, during the process of striking, gathers up the quarter rack, stopping when the rack end has reached the hammer pallet, P_2. The quarter rack, *Q*, has two separate groups of three ratchet teeth which, in returning, alternately come into contact with the hammer pallets, P_1, and P_2, in turn actuating the hammer pins which project through the plate from the opposite side. The hammers themselves, M_1, M_2 (Fig. 97B), act on different gongs to produce the double-tone strike for the quarters only, and are returned to position by means of return springs, R_1, R_2 (Fig. 97A). The quarter snail, *L*, mounted on the cannon pinion is made with four steps to set the quarter rack correctly and, projecting radially below the highest step, a portion of the surprise-piece can be seen. The purpose of the surprise-piece is to avert the possibility of the third quarter being struck again just at the critical moment when the hour changes and for a short while after. It consists of a segment of a disc colleted freely, concentrically with the snail, extending completely below the highest step, and is limited in its motion by a pin acting in a slot. This surprise-piece carries a block for advancing the star, *V*, once an hour, and as this takes place the block, in the effort to propel a tooth of the star, is temporarily taut and its complementary segment remains concealed behind the face of the highest step of the snail. In this position the quarter rack is still free to fall into the deepest part of the snail, but the instant the star wheel jumps, the block, which occupies most of the space between two star teeth, is pushed forward by the next tooth and the surprise-piece moves out to form an extension of the highest step of the snail. There is then no danger of the third quarter being repeated after the hour has passed. As the cannon pinion and snail rotate, the block moves out of the path of the star wheel and the surprise-piece then remains out of action until the next hour approaches.

Returning to the barrel arbor; mounted below the quarter-rack hook-piece or gathering pallet, *G*, and steel winding wheel, *D*, the hammer ratchet, *H*, is partly visible. This is the hour-striking rack carrying 12 teeth over a segment representing about one-third of what would be the total circumference. It is rotated with the barrel arbor and the number of teeth appropriate to the position of the hour snail pass the hammer pallet, P_2, and striking takes place as the rack returns. The hours are thus reproduced on one gong only and the pieces are carefully adjusted so that the hour striking is completed before the gathering pallet on the barrel arbor reaches the quarter rack.

Fig. 97B shows the reverse side of the plate where gongs and hammers M_1 and M_2 are visible as well as the barrel bar and ratchet for locking the back action.

By way of comparison it is interesting here to return to Plate X and inspect certain features of the system adopted in this specimen by Le Roy. In most old repeaters it was the practice to mount the star wheel and hour snail below a long lever riding on a stud at one end. This lever, which can be seen in the front view, is actually the all-or-nothing piece and has a small amount of side play, though normally it is held towards the winding lever by means of a spring.

The mechanism is wound by pressing the pendant upon a lever connected by a chain with the barrel arbor, on which is mounted a circular striking rack. Part of this has 12 teeth to reproduce the hours and the remainder three sets of two teeth each for the quarters. There is only one hammer (see side view) beating a dumb block, so that the two teeth passing in quick succession give rapid double blows. The circular motion afforded to this rack is determined by the position of the hour snail, against which an arm forming part of the pendant lever butts when the pendant is pressed in for winding the striking. The snail yields also to the small amount of side play provided and releases the all-or-nothing piece. By this means a locking lever is set free and a quarter rack of three teeth immediately drops on to the quarter snail mounted on the cannon pinion. Squared on the barrel arbor there is, however, a peculiar and most important cam with three notches, which can be seen in contact with a short pointed arm projecting from the toothed end of the quarter rack. As the mechanism is wound and the striking rack moves into position, this cam moves away from the arm on the quarter rack, but with the setting of the quarter rack after the all-or-nothing piece has released the locking lever, the arm approaches the cam. At the first quarter, all three teeth on the quarter rack pass and the point of the arm is then nearest the centre of the striking rack. As the latter returns, the hours are struck, but, immediately afterwards, the notch nearest the centre of the cam catches the quarter rack arm, driving the rack home and throwing over the locking lever until it is gripped by the end of the all-or-nothing piece. The procedure is similar with the other quarters, but there is another point to notice. Unlike the modern surprise-piece, the purpose in this case is for the projecting segment to keep the quarter rack on the highest step of the snail before the hour, not after the hour. The steps on the snail are reversed, the lowest being the silent quarter instead of the highest. A further feature of interest is the arrangement visible in the back view for regulating at will the speed of the repeating train. The square on the little hook at the top of the illustration can be turned right or left with a watch key, thus bringing the nose of the hook in or out of the path of the wire lever seen pivoted to a cock near the edge of the frame. This lever is mounted on the pallet staff of a recoil escapement, the motion of which is then controlled by the amount of side play of the lever.

Minute Repeaters. Of necessity, the additional apparatus involved in reproducing minutes as well as hours and quarters renders the mechanism considerably more complicated. Fig. 98 illustrates the front plate of an English hunter movement with a minute-repeater mechanism made entirely by the old-established Clerkenwell firm of Usher and Cole. Particular interest is attached to this specimen, because all-English modern repeater mechanisms are extremely rare, those which are usually provided in English movements being Swiss made and only adapted in this country.

Besides the quarter rack, which bears very little resemblance to that used in the quarter-repeater mechanisms, there is another mounted above it on the same stud, known as the minute rack, *K*. At the left-hand extremity of this there are 14 ratchet teeth corresponding with the number of minutes between each quarter. It has also six ratchet teeth which are entered by the rack hook, *L*, known by some as the minute and quarter coupling This piece is attached by means of a shoulder screw to the quarter rack, *Q*, and has a tail which lies in the path of a pin above the hour-hammer pallet, P_1. When the two racks fall, the hammer pin deflects the tail and rack hook clear of the minute rack, which is thus permitted to proceed until its

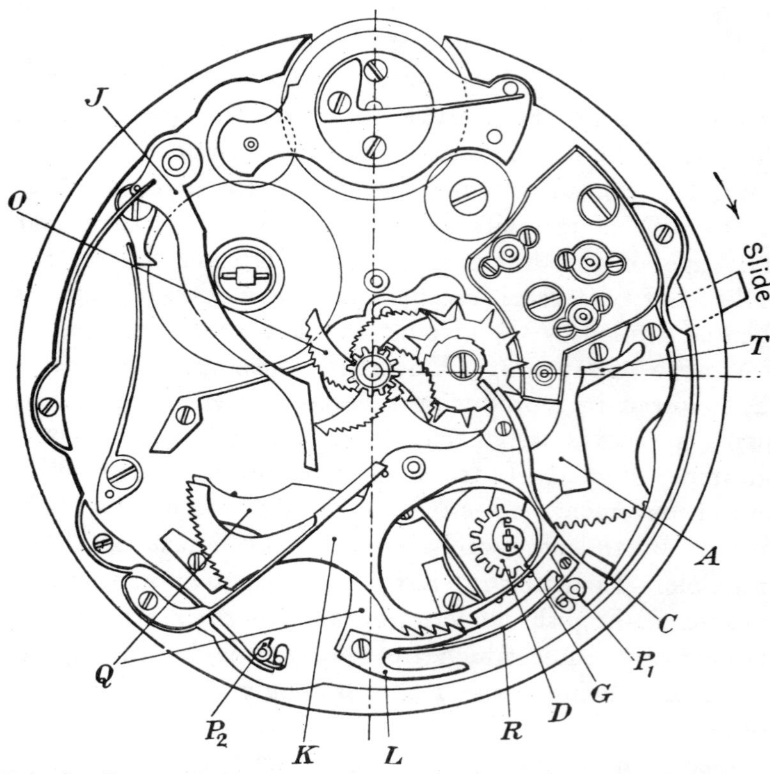

FIG. 98.—Bottom plate of an all-English hunter minute-repeater mechanism.

right-hand extremity or tail reaches the minute snail, *O*, set to the appropriate position by the centre arbor. On the return journey, the rack hook actuated by the spring, *R*, comes into play when the quarters have been completed, and proceeds to gather up the minute rack, the teeth of which operate the hammer pallet, P_2. The minute snail, *O*, comprises four curved wings, each of which is cut with 14 steps. This is mounted above the quarter snail, attached to which, and between the two, is a four-armed surprise-piece to act similarly to the surprise-piece described for use in quarter repeaters, but in this case with a view to keeping back the minute rack on completion of each quarter. A piece known as the quarter jumper, *J*, controls the action of the four-armed surprise-piece and serves a purpose similar to the block and star wheel operation in the case of the quarter surprise-piece. The quarter rack, *Q*, which is mostly concealed in the illustration, has a series of internal teeth besides the two sets of three external ratchet teeth for producing the double blows on the gongs. These internal teeth gear with a steel wheel, *D*, mounted to run freely on the barrel arbor and only partly cut with a corresponding number of teeth to those on the rack. Above this loose steel wheel and squared to the barrel arbor is the quarter-gathering pallet itself, *G*, and as the train runs down and the hours have been struck the gathering pallet meets a pin projecting from the steel wheel below, which it carries round and so gathers up the quarter rack.

The arm, *A*, which acts on the hour snail and by means of the trigger, *T*, removes the all-or-nothing piece from the notch, *C*, in the quarter rack, is partly visible but the hour rack is hidden. The other parts are similar to those as described for quarter repeaters and lettered accordingly.

Fig. 99 shows a front and back view of an up-to-date fine Swiss hunter minute repeater. The parts are set out rather differently from the English mechanism, but the method is really the same.

The principal deviation of plan in this particular calibre (Fig. 99A) is seen in the rack hook, *L*, whose tail projects behind the point where it is mounted on the quarter rack, *Q*. As the racks fall, the rack-hook tail slides against the end of

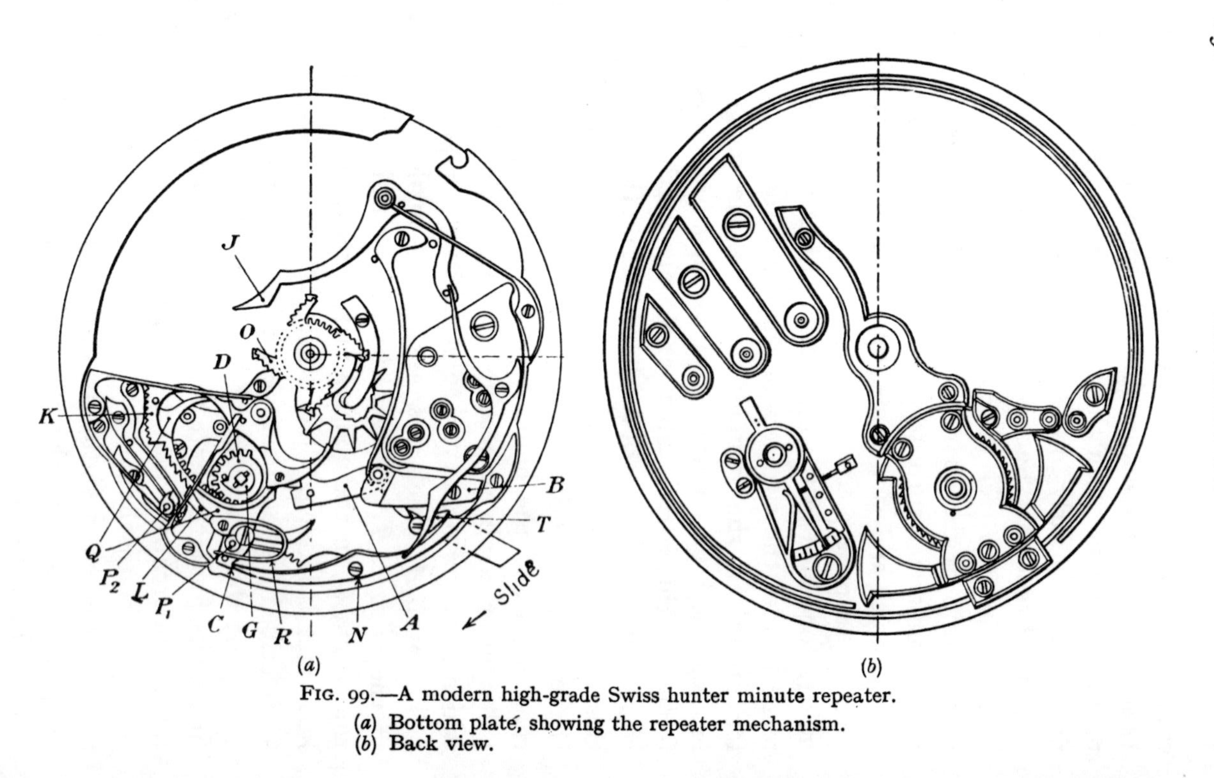

FIG. 99.—A modern high-grade Swiss hunter minute repeater.
(*a*) Bottom plate, showing the repeater mechanism.
(*b*) Back view.

the steel bar, *B*, to which the arm, *A*, is pivoted, thus raising the click end clear of the minute rack, *K*. The quarter rack, in returning, ultimately frees the rack-hook tail from the bar, *B*, and allows the click, by means of the light spring, *R*, to enter the teeth of the minute rack, *K*.

The all-or-nothing piece, *N*, is in this case a spring lever mounted with side play sufficient to allow it to be deflected by the trigger, *T*, when the slide is pushed right home, and so release the quarter-rack tail at the notch, *C*. With the mechanism at rest, the hour-hammer pallet, P_1, can be observed in the figure to be almost completely surrounded by the tail of the quarter rack and rack hook. The minute-hammer pallet can be seen at P_2.

An important feature, in contrast to the English mechanism, shown in Fig. 98, is that the quarter jumper, *J*, is kept out of action, until required, by an arm controlled by the all-or-nothing piece. The extra work put upon the watch in raising the jumper over the minute snail is thus averted. Another method adopted in some designs consists of a special cam mounted on the barrel arbor for the purpose of operating a lever which lifts the jumper off the snail when striking ceases.

Other parts are lettered similarly to those shown in Figs. 97 and 98.

Plate XI is an illustration of a high-grade minute repeater from Mr. Otto's collection. It is interesting inasmuch as the first view shows the mechanism at rest whilst the second depicts it with the slide pulled home in readiness to strike. Unfortunately in the process of reproduction the external portion of the slide has had to be omitted. The arrangement of the different parts can be followed in conjunction with the descriptions already given of other designs.

Half-quarter Repeaters and Five-minute Repeaters. The mechanism is slightly modified in these two varieties, but a detailed description seems hardly necessary as the main features are very similar. In the five-minute repeater, the hammer is operated by an eleven-toothed rack which is arranged to drop upon a snail on the centre arbor having 12 steps. The half-quarter repeater is not quite so simple in arrangement. Similarly to the minute repeater, two racks, though in this case almost identical in form, are mounted one above the other, on the same stud to work together in the same plane. The lower of these two is the usual quarter rack, but the upper is the half-quarter rack and has one tooth only. A pin in the quarter rack projects into a slot in the half-quarter rack, which is thus given additional freedom of movement beyond the quarter rack. Normally, this tooth exactly covers a tooth of the quarter rack, but when pushed over the full extent afforded by the slot the single tooth has advanced in front of the quarter teeth, though retained at one end or the other by a spring claw. The most important feature, however, lies in the fact that there are two snails on the cannon pinion, one for the tail of each rack. The quarter snail has four steps in the ordinary way, but above it the half-quarter snail, also with four steps, is planted so that its steps rise midway between the quarter-snail steps. The effect produced is as follows :—After an hour has been struck, both snails present their highest step to their respective rack tails and then no quarters can be given, but when the quarter step has moved half-way towards its next step the half-quarter snail reaches its first step, so that though the quarter rack is still unable to set for striking the half-quarter rack can now drop forward in advance of the quarter rack, as far as the length of the slot will allow. As it returns, the tooth causes a single blow to be struck and

Plate XI.—The front-plate of a high-grade minute-repeater movement. Left: The repeating mechanism at rest. Right: The slide pulled home in readiness to strike.

(Otto collection.) *Photograph by Mr. W. Beckmann.*

[*To face page 232.*

PLATE XII.—The front-plate of a complicated watch movement. Realising the most wonderful mechanical achievement in watch work, this mechanism combines in one, a clock-watch, minute-repeater, split-seconds chronograph, and a perpetual calendar. Left: The clock-watch, minute-repeater and split-seconds work. Right: The perpetual calendar mounted above.

(*Otto collection.*) *Photograph by Mr. W. Beckmann.*

[*To face page 233.*

thus it proceeds until the next quarter is reached when the edges of the two snails again coincide. The two racks then act as one, because the single tooth of the half-quarter rack is, in this position, exactly covered by one of the quarter-rack teeth. This continues until the next half-quarter, when another step is reached on the half-quarter snail so that independent action is again permitted and, after the double-tone quarter has sounded, a single blow is struck to denote an additional seven and a half minutes. The same procedure recurs throughout the remaining quarters, the half-quarter rack operating conjointly with the quarter over the first half and independently over the second.

Complicated Watches. The highest mechanical achievements do not terminate at repeater mechanisms alone, and there are yet occasional specimens met with which combine many complicated features into one and present the utmost intricacy.

Plates XII and XIII are again views of mechanisms taken from Mr. Otto's collection.

Plate XII shows a wonderful combination comprising a clock-watch, minute repeater, split-seconds chronograph and perpetual calendar. The left-hand view shows the quarter and hour striking mechanism, as well as the minute-repeater and split-seconds work, whilst in the right-hand view the calendar mechanism can be seen, which is superimposed and under the dial.

Plate XIII shows very clearly the perpetual calendar work of another fine complicated watch. The method of advancing the days of the week, the months and the lunar phases can be followed easily but the date work is more involved. In order to provide for the variation between the long and the short months there is a wheel like a locking-plate which revolves once in four years. According to the depth of the slots in this 4-year wheel the position of a connecting lever determines the amount by which the date wheel is advanced on the last day of the month. If the lever rides on the outside edge of the 4-year wheel no supplementary action takes place and the date wheel of 31 teeth is advanced one tooth at a time throughout the entire number. On the other hand if the lever is in the deepest notch then the supplementary action takes place at the 28th tooth and four teeth are gathered up at once and the figure 1 is indicated on the dial. A similar action takes place with the 30-day month. The 4-year wheel, of course, provides for leap-year by having a notch whose depth enables the lever to advance three days in one instead of four at the end of February.

Plate XIII.—The perpetual calendar work from a fine complicated watch. The locking-plate can be seen on the left, which regulates the number of days of the month and in order to include leap-year makes one revolution in four years.

(*Otto collection.*) *Photograph by Mr. W. Beckmann.*

[*To face page 234.*

CHAPTER XIX

EPICYCLIC TRAINS

Tourbillons. In an endeavour to solve the problem of position errors, Bréguet, in 1795, conceived the idea of planting the escapement in a revolving cage or carriage. This carriage was mounted on the fourth pinion and driven by the third wheel. The fourth wheel was a fixture, being screwed to the plate concentrically with the carriage, leaving a hollow centre to allow a passage for the fourth pinion which carried the seconds pivot. The escape wheel was planted eccentrically in the carriage and the pinion was pivoted into a pottence in order to engage with the fixed fourth wheel. Thus, as the carriage was revolved by the third wheel, the escape pinion was propelled by the fourth wheel.

These mechanisms, which Bréguet named "Tourbillons," involved extremely delicate and troublesome construction and were not numerous in his day, but they have until recent years always found a place among the higher grades of precision watches submitted for competitive trials.

Karrusels. In 1894 Bonniksen introduced a more robust form of revolving escapement, shown in Fig. 100, which he designated the "Karrusel" (derived from the Italian *carosello,* meaning "roundabout"). The carriage was not pivoted as in the tourbillon, but had a large hollow brass stem with which to revolve on a bearing in the bottom plate. Below the hollow stem, allowing sufficient endshake, the carriage wheel was screwed concentrically, so that between the carriage and the carriage wheel both sides of the bottom plate also formed large surface bearings. The speed of rotation was much slower,

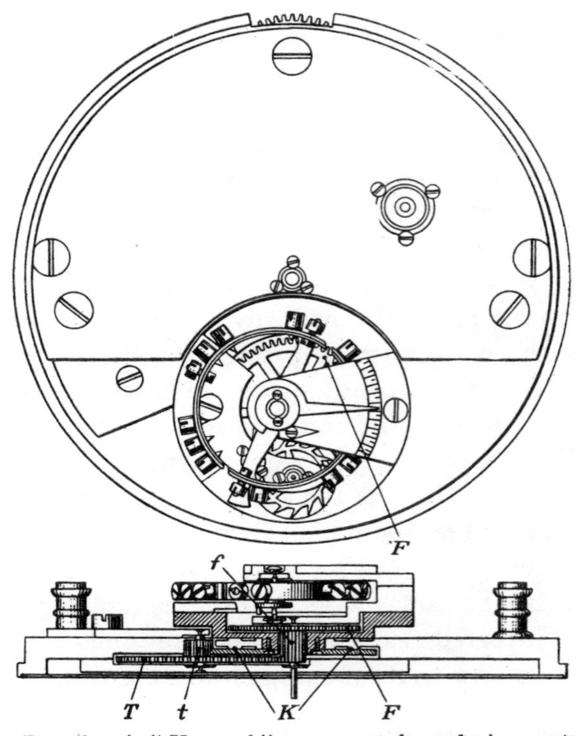

FIG. 100.—Bonniksen's " Karrusel " movement for reducing position errors. The escapement is mounted in the carriage, *K*, which is revolved slowly on a single large brass bearing by the third pinion, *t*, once in about 40 minutes.

generally once in about 40 minutes, and the fourth wheel was not a fixture. The fourth wheel was situated within the carriage, but the fourth pinion passed through the hollow stem to a bar screwed to the bottom plate. The carriage wheel was driven by the third pinion and the third wheel geared into the fourth pinion.

The elevation in the figure shows the carriage wheel, *K*, in section so that the teeth engaging with the third pinion, *t*, cannot be discerned. The third wheel, *T*, mounted on the third pinion, gears with the fourth pinion, *f*, pivoted so as to rotate with the carriage. The fourth wheel and the complete escapement are within the carriage. An important fact to remember is that the effect of one complete turn

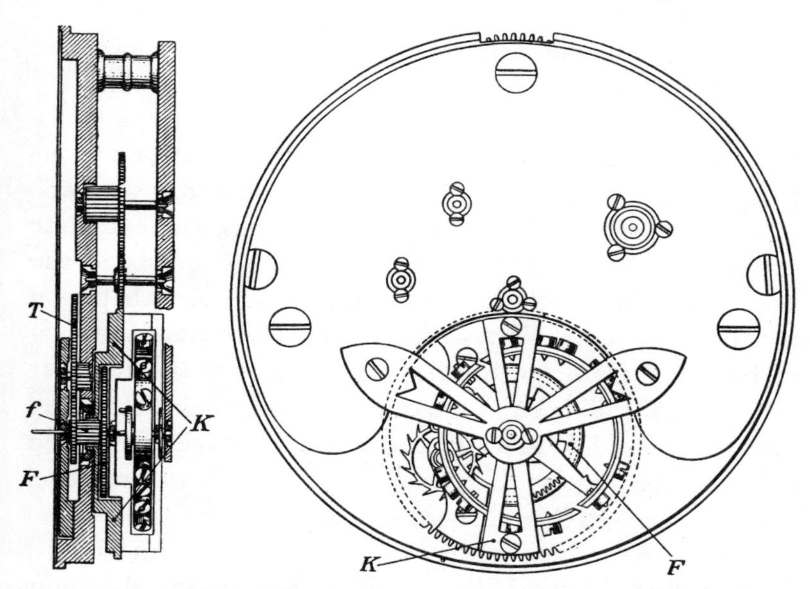

FIG. 101.—Bonniksen's " Tourbillon " movement. The pivoted carriage, *K*, is made to revolve in the reverse direction to the fourth wheel, *F*, once in 39 minutes, by an independent train from the centre wheel.

of the fourth wheel is either lost or gained during one turn of the carriage—lost if the carriage turns in the same direction as the fourth wheel, but gained if the carriage turns in the opposite direction to the fourth wheel.

Bonniksen's "Tourbillon." In 1903 Bonniksen also introduced a form of tourbillon which, like his " Karrusel," was made to revolve slowly. This is shown in Fig. 101. The carriage, K, which is driven by the centre wheel through an extra wheel and pinion and an intermediate wheel, revolves once in 39 minutes and really comprises two compartments. In the bottom it has a large pivot which rides in the bottom plate. The pivot is hollow to allow a free passage for the fourth pinion, *f*. The bottom compartment of the carriage is recessed to receive the fourth wheel, *F*, which is covered by a bar to carry the fourth top jewelling. This bar forms the bottom of the second compartment into which the escapement is fitted. The bottom staff jewelling is eccentric to the top fourth jewelling on the other side of the bar and the balance cock, which covers the second compartment, is provided with a top pivot to work in the two-armed bar screwed to the top plate. Thus, the complete carriage is pivoted top and bottom and so reverts to the usual and safer method of mounting. The fourth pinion is driven by the third wheel, *T*, and the bottom pivoting of both the third and fourth pinions is in a separate bar screwed to the bottom plate. The intermediate wheel referred to which actually drives the carriage is inserted only for the purpose of reversing the direction of rotation of the carriage to that of the fourth wheel.

Besides Bonniksen there were others who introduced different forms of tourbillons about the same time. Amongst these one brought out by Taylor, with an internal gear, was of particularly unique design.

Calculation of Epicyclic Trains. Epicyclic trains are so called from the fact that one of the wheels, being in gear with another, not only rotates about its own centre, but also rotates about the wheel it engages. The tendency thus produced is to increase or decrease the total ratio of the train according to the direction and speed of rotation of the wheel which performs the double function. Certain rules, differing from those described in Chapter XII, therefore, have to be observed in the calculation of karrusel and tourbillon trains.

Expressed as a formula similar to that for ordinary trains, epicyclic trains may be treated as follows: C, T, F, E represent, as before, the numbers of teeth in the centre, third, fourth and escape wheels respectively, and t, f, e, the numbers of teeth in their corresponding pinions. N = the number of revolutions of the carriage per hour. The sign between the two terms of the formula is plus or minus according to whether the direction of rotation of the carriage is contrary to or the same as that of the fourth wheel.

$$\text{Vibs. of balance per hour} = \frac{2CTFE}{tfe} \pm \frac{2NFE}{e}.$$

To meet the requirements of a seconds train the formula can be modified, because then $\frac{CT}{tf} = 60$ and, with the usual 15-tooth escape wheel, the expression by substituting will become

$$\text{Vibs. per hour} = \frac{1800F \pm 30NF}{e}.$$

The train generally employed in karrusel watches is 18,000, which equals five beats to the second, and the equation resolves itself into :—

$$1800F \pm 30NF = 18{,}000e,$$
$$60F \pm NF = 600e.$$

Similarly a 14,400 train, beating four to a second can be reduced to the simple formula :—

$$60F \pm NF = 480e.$$

The carriage is usually driven by the third pinion, though sometimes it is through a separate head on the third arbor.

In the former case the carriage pinion, k, is the same as t, so that :—

$$N = \frac{C}{K},$$

but, in the second case :—

$$N = \frac{Ck}{tK}.$$

If, as in most instances, $\frac{C}{t} = 8$,

then $$N = \frac{8k}{K},$$

or $$K = \frac{8k}{N}.$$

Example. To illustrate the application of the preceding formulæ, it is interesting to consider the possibility of constructing a karrusel with an 18,000 train and a carriage making 1 turn per hour, rotating, say, in the opposite direction to F. The sign is + and $N = 1$.

Then $$60F + F = 600e$$

and $$61F = 600e$$

$$\frac{F}{e} = \frac{600}{61}.$$

It is immediately apparent that such a solution is impracticable. 61 will not split into factors, and it cannot be the number for an escape pinion. If the carriage turned in the same direction as F the escape pinion would be 59, which is just as bad.

The obvious number for the escape pinion would be either 7 or 8, so that taking the first instance and assuming still that $N = 1$ with a rotation in the same direction as F, then

$$F = \frac{600 \times 7}{59} = 71{\cdot}18.$$

But supposing the carriage makes 2 rotations in the hour, then $N = 2$, and

$$F = \frac{600 \times 7}{(60 - 2)} = 72{\cdot}4.$$

It follows, of course, that neither of these results is the slightest use, as the fourth wheel must have an equal number of teeth and the carriage, therefore, cannot be arranged to rotate once in either an hour or half-an-hour, whilst an escape pinion of 7 is used. But if 72, which comes between these two numbers, is chosen as the number for the fourth wheel then the time of rotation of the carriage can be ascertained. The formula :—

$$60F - NF = 600e$$

becomes $$(60 \times 72) - 72N = 600 \times 7$$

or $$4320 - 72N = 4200,$$

whence $$-72N = -4320 + 4200$$

and $$N = 1\tfrac{2}{3}.$$

That is, N makes $1\frac{2}{3}$ turns per hour or 1 turn in 36 minutes.

Similarly, it may be shown that if an escape pinion of 8 is used, the fourth wheel requires to be 82 and one rotation of the carriage is then made in 41 minutes.

By using now the formula, $N = \frac{8k}{K}$ and assuming that k, being one and the same with $t = 10$, then the number of teeth in the carriage wheel can be ascertained as follows :—

$$\frac{5}{3} = \frac{8 \times 10}{K}$$

$$K = \frac{240}{5}$$

$$= 48.$$

It is thus possible to complete the entire train and similarly calculate values under different circumstances.

The attempt to eliminate position errors by means of these devices showed conclusively that the effect of the eccentric centre of gravity upon timing (discussed in Chapter XIII) was not known to those who invented and adopted the principle of the revolving escapement. It was certainly not understood in Bréguet's time, but now it is so carefully studied in the springing that the best possible timing results from watches with fixed escapements. Only as far as English work is concerned have the rates of watches with a fixed escapement been improved upon by tourbillons and karrusels. At the Kew Observatory Trials the highest marks obtained have been 93·9 and 92·7 for English tourbillons and karrusels respectively. With fixed escapements, however, the best result for an English watch was 91·2 in 1909, whilst among Swiss watches competition has grown enormously in recent years and the remarkable record of 97·00 marks was reached in 1923 and 97·2 in 1926.

Thus, in the light of present attainments it cannot be maintained that watches with revolving escapements show any superiority whatever to correctly-planned fixed escapements and theoretical springing. The delicacy and cramping of the mechanism are detrimental to solidity of construction, and faults are consequently introduced, whilst the origin of the revolving escapement can only have been due to a misconception of the fundamental principles of timing. It should be remembered that the position "pendant down" is never taken into consideration at Observatory trials.

Plate XIV.—John Harrison's marine timepiece, No. 4. This is a view of the top-plate and case of the historic masterpiece which won the award of £20,000 offered by the Government for an instrument by means of which longitude could be determined at sea.

By courtesy of the Astronomer Royal, Royal Observatory, Greenwich.
(*Photograph by Mr. F. Jeffries.*)

[*To face page 243.*

PART IV—MARINE CHRONOMETERS

CHAPTER XX

CHRONOMETER TRAINS, ESCAPEMENTS, BALANCES AND OTHER DATA

Introductory. It would scarcely be possible to find any instrument of greater beauty and refinement than the modern box chronometer. The necessary accuracy of performance, demanding the highest skill and craftsmanship, undoubtedly makes it the pride of horological attainment.

The birth of the marine chronometer arose from the necessity of having accurate time at sea, so that, when out of sight of land, it might be possible to determine the ship's longitude. This proved to be a most perplexing problem to the horologists of the time, but a solution was ultimately arrived at by John Harrison, the inventor of the gridiron compensation for pendulums (see p. 29). His fourth marine timepiece, finished in 1759, won, after some controversy, the award of £20,000 offered by the Government for an instrument capable of enabling a navigator to determine longitude within half a degree. Actually, Harrison's instrument surpassed these requirements. Plate XIV shows a view of the back and case of this historic masterpiece, very nearly full size.

From a mechanical standpoint, the problem of automatic compensation for variation of temperature was the principal obstacle to progress. Harrison's method of meeting this difficulty was by using what became known as a "compensation curb." This was a slender arm composed of a strip of brass and another of steel riveted together, so that, with the fluctuations of temperature the free end, carrying two curb pins, moved along the balance-spring near the stud, thus altering the effective length of the spring.

Experimental work along similar lines was continued by Harrison's contemporaries, but their efforts resulted in very little of a lasting character in the evolution of the marine chronometer. Mudge, nevertheless, devised numerous ingenious and original contrivances, notably his remarkable remontoire or constant-force escapement. It should be borne in mind that these early makers were all imbued with the idea that the secret of accuracy of timekeeping lay principally in the providing of uniform impelling power. It was this fact that led to the many more or less complicated efforts along the lines of remontoires.

The greatest progress, however, towards the instrument of to-day was made by Pierre Le Roy (1717–85). His father, Julien Le Roy, is referred to in Chapter XVIII, as having introduced the "all-or-nothing" piece and other improvements and refinements to striking watches. To Pierre Le Roy credit is due for originating two of the most important features which have brought the marine chronometer to its present state of perfection. One was the detached escapement and the other the compensation balance. He realised the importance of isolating the balance and balance-spring as far as possible from any interference of the train and also perceived the virtual necessity of ensuring what is now known to be isochronism. Instead of altering the effective length of the balance-spring by means of a compensation curb, Le Roy proved that the most desirable practical method of equalising the effect of fluctuating temperature was through the balance which he constructed in the form of a rim of brass and steel laminæ riveted together.

Ferdinand Berthoud (1727–1807) was another pioneer whose work must not be passed unnoticed. He was a contemporary and rival of Pierre Le Roy, and his experimental work covered an extraordinarily wide area. In spite of his many fine productions, however, he left nothing of outstanding merit which has been passed on to the modern instrument.

The development in this country advanced more rapidly at the hands of the two renowned horologists, John Arnold (1736–99) and Thomas Earnshaw (1749–1829). Arnold concentrated his efforts on improving the compensation balance and the balance-spring, besides introducing a form of spring-detent escapement, which contrasted with the former pivoted detent escapements originated by Berthoud.

Arnold's accomplishments were nevertheless eclipsed by those of Earnshaw, who not only devised the spring-detent escapement which has survived practically unchanged to the present day, but also solved the difficulty of unifying the two metals used in the compensation balance by fusing them together instead of riveting them or soldering them as Arnold had done. Earnshaw's practice was to turn a channel in a steel disc forming the body of the balance. Into the channel the brass was fused and the superfluous metal turned down to the correct dimensions. The steel disc was crossed out leaving a diametrical crossbar, whilst two adjustable brass weights were added to the rim. About one-eighth of the whole rim was cut away at each end and on opposite sides of the crossbar, and the weights were mounted to suit the compensation necessary. Timing screws were inserted in the rim at both ends of the crossbar. This principle of construction is yet in force and the actual form of balance used to-day bears a very close relationship to that introduced by Earnshaw (see Fig. 104).

From this brief *résumé* of the pioneering days it will be noticed that, apart from Harrison, who made the first successful machine for determining longitude at sea, the improvements by Pierre Le Roy and Thomas Earnshaw created the essential foundation upon which the modern marine chronometer is based. For a complete history of the evolution of these instruments, the reader may be referred to Lt.-Comdr. Rupert T. Gould's erudite work, *The Marine Chronometer, its History and Development* (Potter, London). In this will be found a perfect repositorium of interesting facts on the subject, written in the most acceptable form possible.

Movement. Passing now to the modern marine chronometer, the principal features to be considered are the escapement and the balance and balance-spring. The movements are almost invariably constructed to run for two days and the train is made up as follows :—

The fusee gives 8¾ turns.
The great and centre wheels have 90 teeth.
The third and fourth wheels 80.
Centre pinion 14.
Third pinion 12.
Fourth and escape pinions 10.

The train is thus arranged so that the balance beats 14,400 vibrations per hour or four beats to the second and, as the escape wheel only advances with alternate beats, the seconds hand moves forward every half second. Viewed from above, the top-plate, Fig. 102, shows the arrangement of the train, which resembles closely that of the original English full-plate lever watch on a larger scale. The balance staff is pivoted between a cock and pottence screwed above and below the top-plate. The fourth pinion is provided with a long bottom pivot passing through the dial to carry a seconds hand and the bottom fusee pivot operates an up-and-down work which moves a hand over

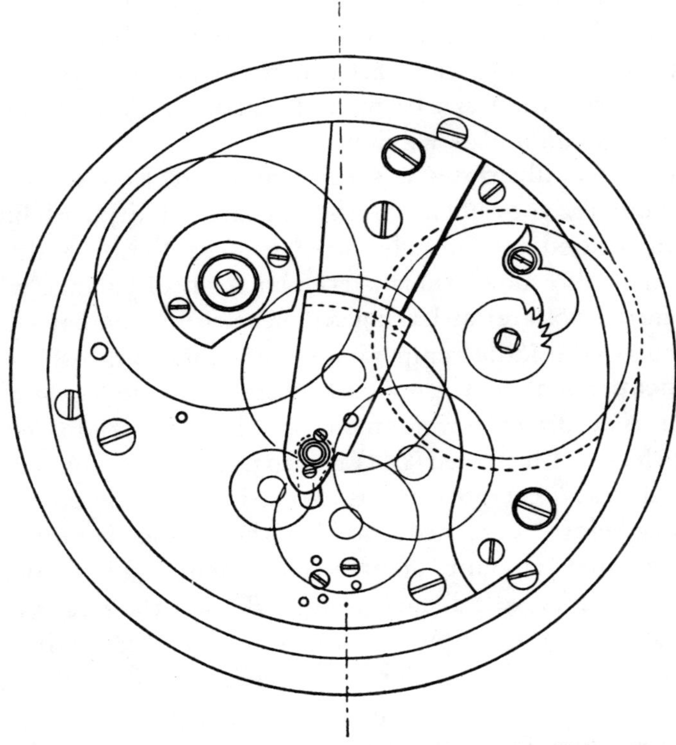

FIG. 102.—Illustrating the calibre of a modern Marine Chronometer movement.

an appropriate segment on the dial. The complete movement is fitted into a heavy brass box and the bezil and glass are screwed down to hold it firmly. A large pipe is screwed to the top-plate over the winding arbor for the purpose of protecting the balance, and extends to a hole in the brass box for inserting the key.

A specially adapted wooden box is used to accommodate the chronometer and, in order to maintain the instrument in a horizontal position, gymballing is provided. This consists of a brass ring pivoted to the wooden box. The brass box is pivoted inside this ring in such a way that the two move about diameters at right angles to each other. The wooden box is made in two sections so that the lower portion containing the chronometer can be kept locked, whilst observation can be made by lifting a lid and looking down through a glass window.

Plate XV illustrates a specimen of an English marine chronometer movement of the highest quality and finish. Manufactured by the celebrated firm of Victor Kullberg, this view also shows their particular form of balance which is referred to later as having so successfully combatted the difficulty of middle-temperature compensation.

Escapement. As previously stated, the modern escapement is a direct development of the one originated by Earnshaw. Illustrated in Fig. 103, it comes under the "dead-beat" category and is also a single-beat escapement, besides being "detached" so that the balance is permitted a maximum amount of free movement. The balance-staff carries large and small rollers, R_1, R_2, the larger being provided with an impulse jewel, I, which receives impulse from the wheel. A slender steel spring-detent, D, is planted so that its free end very closely approaches the small roller. This detent carries the locking stone, L, and also has screwed to it lengthwise a thin gold-spring, G, or passing-spring, whose extremity projects over the end of the detent itself into the path of the discharging pallet, E. This is mounted in the small roller, R_2, which can be adjusted to suit the angle between the discharging and impulse pallets. The figure shows a tooth, b, of the escape wheel just leaving the impulse pallet whilst the tooth, a, in

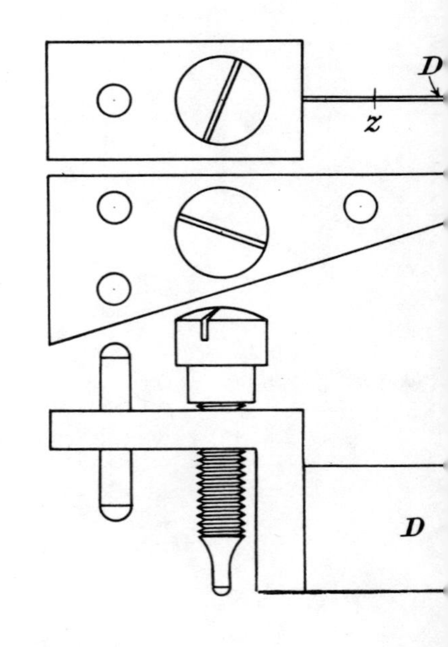
D
z
D

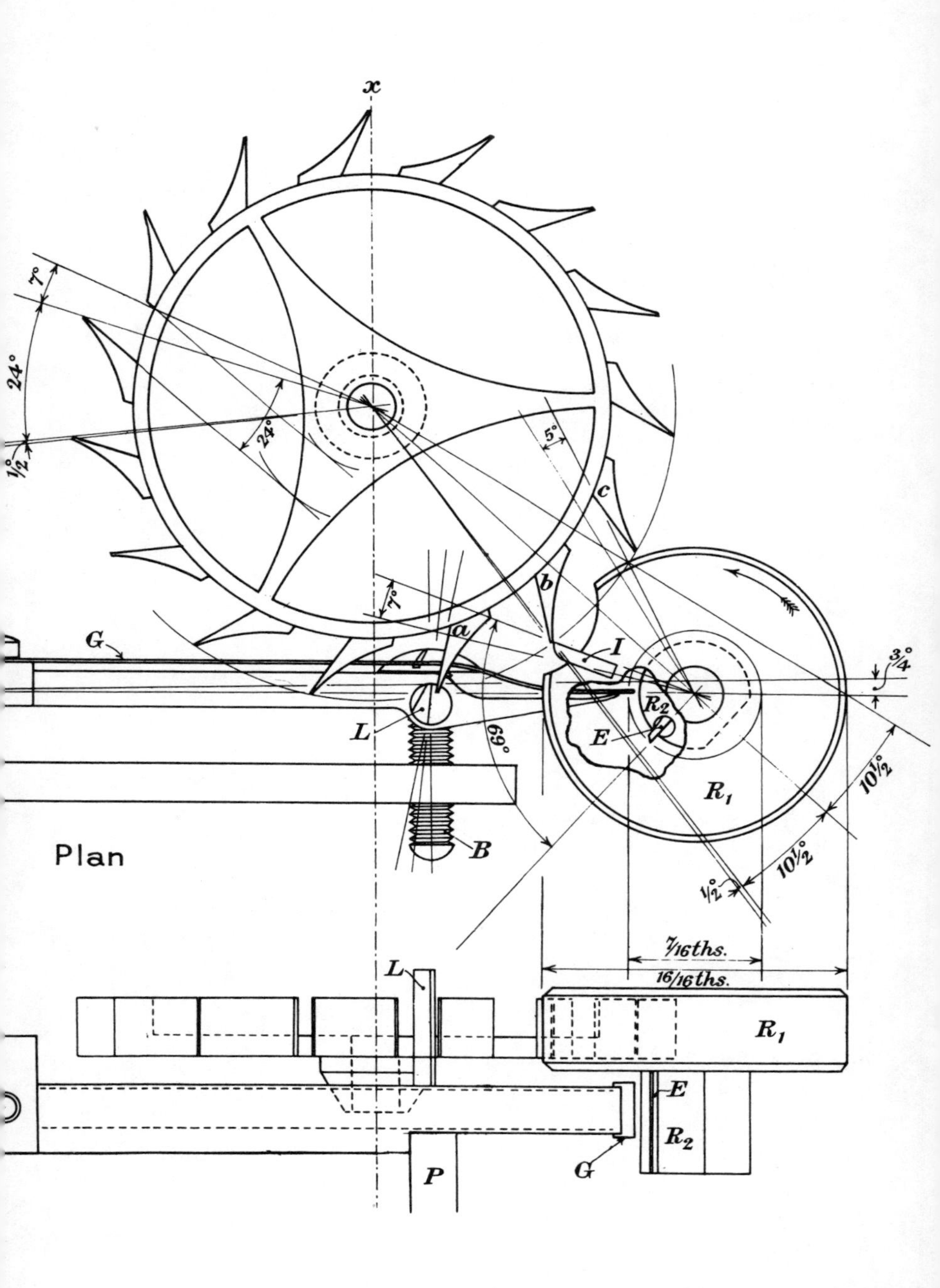

;. 103.—Marine Chronometer escapement. Although credit is due to Pierre Le Roy for originating the spring-detent escapement, this illustrates the modern development of the improved form of construction devised by Thomas Earnshaw.

[*To face page* 249.

front is about to be locked on the stone, *L*. The balance, having received impulse, continues its motion in the direction of the arrow, but on its return it covers the whole vibration unimpeded except that the discharging pallet, *E*, has to pass the gold-spring. On the succeeding vibration, the discharging pallet depresses the gold-spring in passing, and with it the detent, so that the tooth is taken off the locking. At this moment, the impulse pallet has entered the wheel through an angle of 5° measured from the roller centre in readiness to receive the next tooth, *c*, of the wheel. A banking-screw, *B*, is provided for adjusting the depth of the locking correctly and the detent carries a little steel pipe, *P*, to accommodate the locking stone. The diameter of the jewel is equal to about one-third of the distance between two teeth.

The proportions shown in the figure are as follows: The intersection of the wheel circumference with the impulse roller is 21°, that is, the distance between two wheel teeth, 24°, less 3° to allow for the top of the tooth, clearance and drop. The usual actual diameter of a 2-day chronometer escape wheel is 0·56 inch. 24° is allowed for undercut of the wheel teeth, ½° for the width of the tooth and 7° between the face of one tooth and the back of the preceding one. As shown in the elevation, the teeth are raised above the face of the wheel to give lightness to the wheel and strength to the teeth themselves where it is most needed. The acting diameter of the impulse roller is half that of the wheel, whilst the acting diameter of the discharging roller is 7/16ths that of the impulse roller. The angle between the faces of the impulse and discharging pallets is 69°, allowing for the drop of 5° previously mentioned. The width of the impulse pallet covers approximately an angle of 7° measured from the centre of the roller.

In the figure the length of the detent from the line of centres, *xy*, to the block is equal to the diameter of the escape wheel. The length of the spring is one-third of the diameter of the wheel and the bending point, *z*, one-third of the length of the spring from the block. An angle of $\frac{3}{4}°$, measured above the line of centres from *z*, gives the top of the locking stone, *L*. The distance of the acting surface of the locking stone from the tooth, *a*, as shown, is equal to one-half the drop between the tooth, *c*, and the point where the wheel intersects the roller. The acting surface of the locking stone passes diametrically through the centre of the pipe (which, as already stated, is about one-third of the distance between two teeth) and is inclined so as to bisect the angle formed by a line parallel to the line of centres and another parallel to the face of the wheel tooth, *a*.

Balances and Balance-springs. The theories relating to balances and balance-springs have already been discussed in Chapter XIII, and the same rules apply to those used in chronometers. As previously stated, Earnshaw's invention forms the basis upon which the construction of the modern compensation balance is founded. Fig. 104 illustrates the two side by side and their similarity, in spite of the long intervening period, will be noticed at once. In the modern balance, the screws at the extremities of the crossbar carrying heavy nuts are the timing screws, whilst the circular weights situated on the rim are held in place by pinching screws so that they can be moved easily one way or the other if required for purposes of compensation. Two extra weight screws are also shown in the figure.

Marine chronometers are, however, required to function within close limits over a very wide range of temperature. The tests to which these instruments are submitted at the Royal Observatory, Greenwich, extend from about 45° to 100° F., so that the problem of middle-temperature error

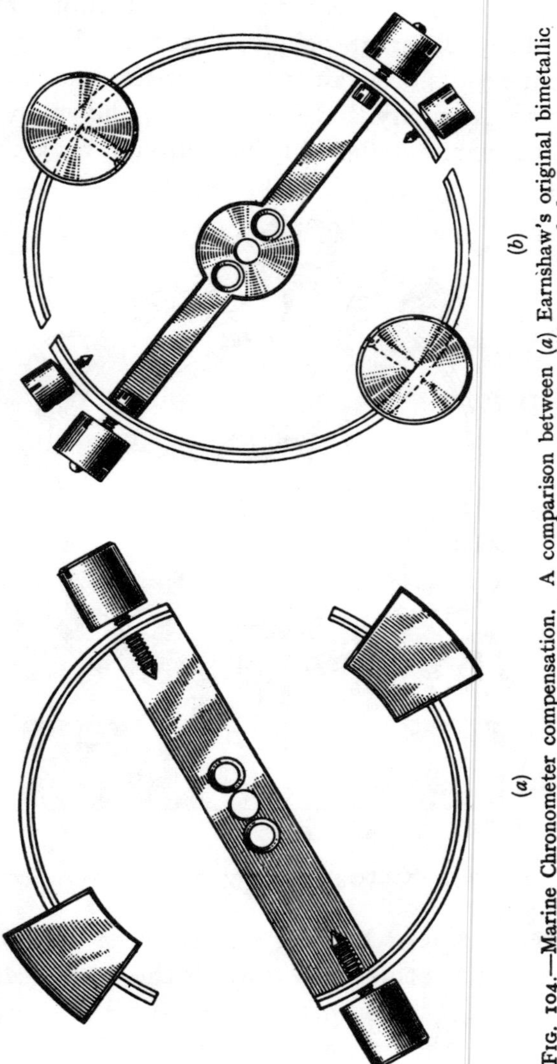

FIG. 104.—Marine Chronometer compensation. A comparison between (*a*) Earnshaw's original bimetallic balance and (*b*) the modern plain brass and steel bimetallic chronometer balance.

has always proved a serious obstacle to manufacturers. Many devices for overcoming this trouble have been suggested by different people, but none has given more consistent results or attained the merit of one known as the "Kullberg Auxiliary." This balance, shown in Fig. 105, has the usual bimetallic rim, but through an arc of 80° on

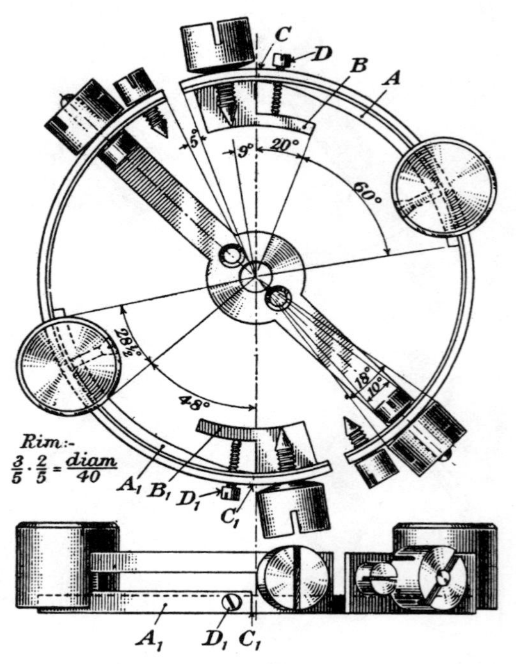

FIG. 105.—Kullberg's auxiliary compensation balance.

the opposite sides, covering one-third of the width, the steel in the rim is thickened towards the centre (A, A_1, plan and elevation). The middle third of the width is left open through the arc, and two blocks B, B_1 are screwed to the main part of the rim. This auxiliary rim, A, A_1, is cut through at C, C_1, and is prevented from inward movement by the screws, D, D_1, banking against the blocks,

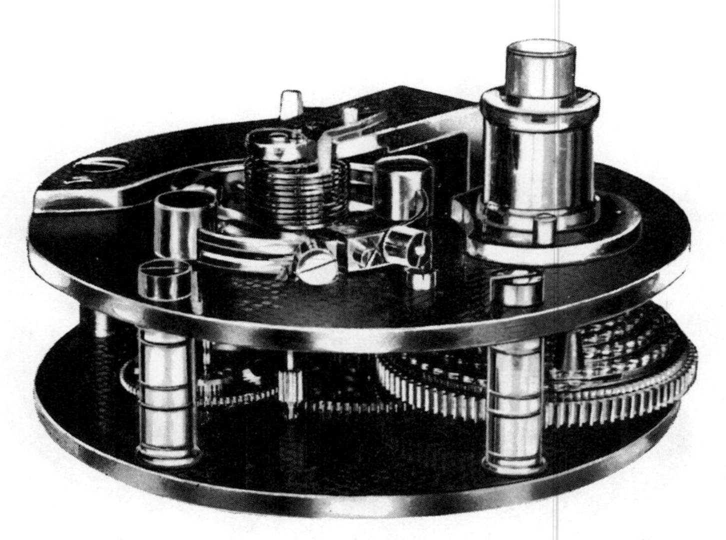

PLATE XV.—An English 2-day marine chronometer movement of the highest grade by Messrs. Kullberg, showing the "Kullberg auxiliary" compensation balance and palladium alloy balance-spring.

[*To face page 252.*

PLATE XVI.—A typical high-grade Swiss 2-day marine chronometer movement by Messrs. Ulysse Nardin. This chronometer is fitted with a "Guillaume" compensation balance.

[*To face page 253.*

B, B_1. The chronometer illustrated on Plate XV is fitted with one of these balances.

Auxiliaries only come into operation under the specific conditions indicated, and otherwise the whole behaves as an ordinary compensation balance.

A famous balance composed of brass and nickel-steel, the invention of Dr. Guillaume, is shown in Fig. 106. Certain combinations of this alloy have produced the remarkable

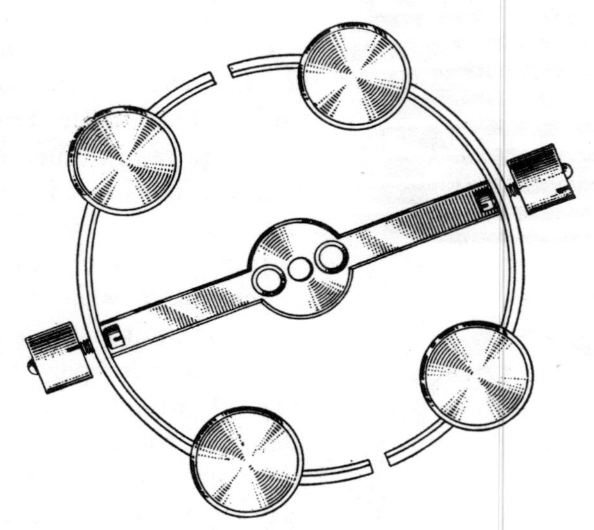

FIG. 106.—Brass and nickel-steel bimetallic chronometer balance, invented by Dr. C. E. Guillaume.

effect of practically eliminating middle-temperature error, and as it is possible with these balances to revert to an original practice of cutting the rim midway between the ends of the crossbar, the centrifugal error is very considerably reduced. Statistics show that, with an ordinary marine balance making about 1¼ turns, the centrifugal error causes a loss of as much as 12 seconds a day, whilst in a balance cut midway between the arms the loss only approximates 2 seconds a day. (See Plate XVI.)

The form of balance-spring used in marine chronometers is invariably helical, as shown in Fig. 107. Generally, it consists of 12 turns or less and is either steel or palladium alloy, though elinvar seems likely soon to occupy a very prominent position.

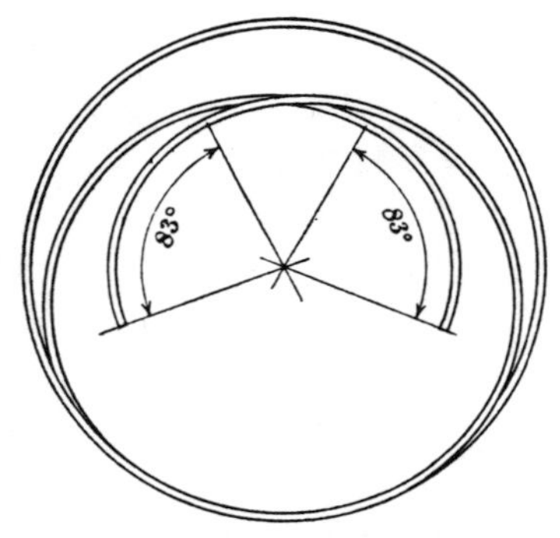

Fig. 107.—Marine chronometer helical balance-spring.

Palladium springs vary somewhat in toughness, some being very brittle whilst others are very soft, and it is a custom amongst springers to age them artificially by raising them to the temperature required in blueing steel. It is a curious fact in connection with this alloy that two springs of exactly the same dimensions may show a difference in timing of as much as 7 seconds per minute. Though dimensions may be identical, the elasticity of the different mixings of the alloy may show considerable variation.

The overcoil shown in Fig. 107 is Lossier's development of Grossmann's construction similar to that of Fig. 72 (p. 166) for flat springs. The theoretical factors are just the same for helical as for flat springs, in order that isochronal conditions may prevail. The upper and lower overcoil may commence at any part of the circumference without interfering with the essential conditions, but for the sake of symmetrical appearance they generally start at points equidistant from the line of centres.

Experimental work is still in progress with regard to the application of elinvar balance-springs to marine chronometers. Monometallic uncut balances have been devised on the Continent by Ch. Volet and Paul Ditisheim, that of the latter, which is on similar lines to the one applied to watches, having given the better results. This achievement still further signalises Mr. Paul Ditisheim's exceptional ability both as scientist and horologist.

In this country also, contemporary work has been most successfully carried out by the firm of T. and F. Mercer. Plate XVII illustrates one of their standard 2-day movements to which they have applied a nickel-steel monometallic balance with short brass compensation curbs of their own special design and a nickel-steel alloy balance-spring. This instrument achieved exceptional merit on being subjected to an unusually rigorous temperature test carried out at the National Physical Laboratory. Throughout the trial, which extended over 11 periods of 5 days each, covering a range of temperature from $-20°$ C. to $+50°$ C., the difference shown between the highest and lowest mean rate for each period was only 7 seconds.

Dr. Guillaume's remarkable discovery has, therefore, given rise to new developments of such importance as to create a fresh epoch in the art of timing marine chronometers.

Trials. For a period of 74 years, until 1914, it was the custom for Greenwich Observatory to hold annual trials of box chronometers, each lasting 29 weeks. The method of obtaining order of merit, the award of what is known as the "Trial Number," of those submitted was based upon a formula $(a + 2b)$, where a is the algebraic difference between the greatest and least weekly sums of daily rates and b is the greatest difference between the sums of daily rates for two consecutive weeks. This system was evolved from a precise investigation of the requirements of the Navy, and the tests included two periods of four weeks each in the oven, where the temperature ranged from about 75° to 100° F., as well as three periods of seven weeks each in the room. There were no ice-box tests. The average trial number of chronometers taking first place over a period of ten years, including and preceding 1914, was 13·7, the most successful during that time being 9·9, in the year 1913.

Surveying Chronometers and Chronographs. With the advance of science, further demands have been made upon the horologists of to-day for the production of precision time-keeping instruments adapted for determining longitude on land or numerous other purposes, and of a standard comparable with that of the marine chronometer.

The Mercer surveying chronometer is an instrument which should be mentioned as representative of this class. The usual 2-day marine movement is used, but electric contacts are provided to give a break circuit at intervals of either a half or one second, or multiples as required. It is thus possible by connecting up to headphones or a megaphone for a remote observer to receive audible and accurate time signals. The mechanism can be started or stopped on the operation of a slide from outside the case and the whole instrument is protected by a stout leather outer case.

Supplementary to the surveying chronometer, a chronograph has been specially designed for use in geodetic survey or wherever precision time-recording becomes an essential feature. In this instrument, which is illustrated on Plate XVIII, a long brass chart drum mounted on ball bearings and weight-driven, is arranged to rotate at either one or two revolutions per minute as required. A traversing gear conveys longitudinally across the drum a carriage upon which are mounted two electro-magnets. The armatures of these electro-magnets are projected towards the drum

PLATE XVII.—An English 2-day marine chronometer movement by Messrs. Mercer. This chronometer is fitted with a "Mercer" monometallic balance and a nickel-steel alloy balance-spring.

[To face page 256.

PLATE XVIII.—A chronograph used in conjunction with the "Mercer" surveying chronometer for measuring small intervals of time.

PLATE XIX.—Another form of chronograph to that shown on Plate XVIII. In this case the record is made on a moving tape instead of upon a drum chart.

[*To face page 257.*

chart and terminate in bucket pens which are deflected when contact is made and their respective circuits closed. One of the pens is wired in circuit with a surveying chronometer, which delivers its signals at regular intervals, whilst the other pen is in circuit with a tapping key operated at will by the observer according to the purpose required. If the object be to determine longitude, the observer would check the chronometer by working in conjunction with wireless time-signals, tapping them on the chart as they are received. It is claimed that by this method an accuracy can be achieved to within 10 yards of longitude.

Another form of chronograph on very similar lines is shown on Plate XIX, the principal contrast being that instead of having a drum chart the signals are recorded on a moving tape.

From the brief reference to these high-grade precision instruments which represent the wider development of the essentially skilled and tested art of marine chronometer production in the past, it will be realised that there is still scope for the advancement of horological attainment to meet the needs of science from its other many and varied aspects.

INDEX

SUPPLEMENT

NOTE

MANY kind expressions of appreciation have encouraged me to add the following pages as a Supplement to the first edition of this book. It is hoped, thereby, to extend its usefulness and to amplify some of the original features.

The reader's attention is drawn to the comments on certain subjects previously dealt with, arising partly from helpful suggestions and also from the fact that further information may make the matter clearer.

J. ERIC HASWELL.

Time Measurement.

(*See pages* 12, 13.)

The fact that certain types of astronomical clocks have attained in recent years such a high degree of accuracy, even in terms of thousandths of a second, makes it necessary to review the more critical features of sidereal time. The slight irregularities in the precession of the equinoxes are known as "nutation," and these are brought about by a combination of different effects of the sun and the moon upon the earth's rotation, which is no longer believed to be constant. Sidereal days are not, therefore, of equal length, and for purposes of adjustment a "mean equinox of date" is imagined as a point proceeding along the equator at a uniform rate. The difference in R.A. between the true equinox and the mean is described as "nutation in right ascension," and may vary between $\pm 1{\cdot}2$ seconds over a period of 18 years.

Sidereal Time is thus to be considered from two aspects—namely, "true sidereal time" as determined by transit observations of the stars, and "uniform sidereal time" as recorded by the sidereal clock, which is equal to true sidereal time minus nutation in R.A. The *Nautical Almanac* has been altered in form to include these functions to three places of decimals. The daily values for nutation in R.A. are shown in the "Sun Tables" and the corresponding values for true sidereal time with the aid of which uniform sidereal time can be ascertained. If the nutation in R.A. sign is minus, the value is added to true sidereal time.

From the foregoing it will be noticed that an analogy exists between true sidereal time and apparent solar time, as determined by observation, and uniform sidereal time and mean solar time, as recorded by clocks, the difference being nutation in right ascension in the one case and equation of time in the other.

Time Checking by Observation.

(*See pages* 18, 19.)

Following example (*a*) for the determination of altitude an allowance is made for refraction. This setting requires to be added and not deducted from the altitude, the object being raised apparently in the heavens (Fig. 4), which produces the resultant value of 69° 7′ 11″ in place of 69° 6′ 29″.

In example (*b*) the correction for the observer's longitude —namely, 20° E.—should be introduced into the final part of the calculation after the G.M.T. of the transit at Greenwich has been ascertained, and the wording should be altered to read thus :

	H.	M.	S.	
R.A. of star	14	11	50	(*i.e.* S.T. measurement eastward of ♈).
*S.T. at Greenwich Mean Noon .	5	21	54	(*i.e.* ♈ at G.M. Noon).
Transit of star at Greenwich . .	8	49	56	after Mean Noon in terms of Sidereal Time.

S.T.	G.M.T.					
8 Hrs. =	7	58	41·36			
49 Mins. =		48	51·97			
56 Secs. =			55·85			
				8	48	29·18 G.M.T.
Less correction for 20° E.				1	20	—
Time of transit at the place of observation .				7	28	29·18 G.M.T.

* N.B.—In the original text the symbol ♈, denoting the first point of Aries, was inadvertently omitted following the parenthesis, thus : " S.T. at Greenwich, (♈ at Greenwich mean noon) . . ."

Early Clocks.

(*See page* 22.)

In recent years doubts have been raised concerning the authenticity of some of the alleged origins of mechanical clocks. The use of the word "horologium," upon which many former ideas seem to have been based, does not necessarily signify anything more than a sundial. This is now considered to be probable in the case of the Glastonbury horologium by Peter Lightfoot, and distinct from the Wells clock, which is thought to have been constructed about 1392. A similar striking clock to the Wells, though without a quarter chime, was made for Salisbury Cathedral a few years earlier. It seems probable that the earliest mechanical clocks emanated from Italy during the first half of the fourteenth century, Milan possessing possibly the oldest relic.

Pendulums.

(*See page* 26, *paragraph* 2.)

In view of the fact that compensated pendulums are also compound, it is more precise to consider pendulums as falling into two classes, Simple and Compound, the latter subdividing into the categories Uncompensated and Compensated.

Dead-beat Escapements.

(*See page* 48.)

Supplementary to the principles of the "Graham" escapement embodied in the description accompanying

Fig. 21, attention should be directed to certain commendable variations of design more particularly concerning its application to large clocks. In the construction of Astronomical Regulators, with compensation seconds pendulums (see page 72), experience shows that the best results are obtained from using longer pallets covering 10½ teeth. In this way it is possible to reduce the amount of drop to a minimum, and the extra friction imposed on the pallet faces is more than counterbalanced by the greater leverage of the longer arms. Furthermore, with the heavy pendulum of a "Graham" regulator, some even weighing 25 lbs., there is always sufficient travel in completing the vibration to permit safe locking within very small limits. A vital feature in such instances is, however, the necessity of providing the most rigid fixing possible for the clock, and the heavier the pendulum the firmer must be the support.

A practical and simple method of setting out the escapement for a regulator, as adopted by the most experienced craftsmen, is shown in Fig. 108. The wheel diameter is usually 2⅛″, and this is equal to the distance of centres between the wheel and the pallets. Centres are very easily obtained by drawing the three circles *a*, *b*, and *c* of equal radius on a straight line. The intersection of the circumference of *a* representing the wheel, being the centre of *b*, and the intersection of the circumference of *b* remote from *a* being that of *c* which is the pallet centre. The intersection of the circumferences of circles *a* and *b* determines the pallet path circle drawn from the pallet centre, locking being on the exit pallet. From the pallet centre ½° is allowed for locking and 1½° for impulse, whilst the length of the impulse faces from the wheel centre is 5°, together with ½° for the tip of the tooth and ½° for drop, making, in all, the 6° which the half space between two teeth of a wheel of thirty permits.

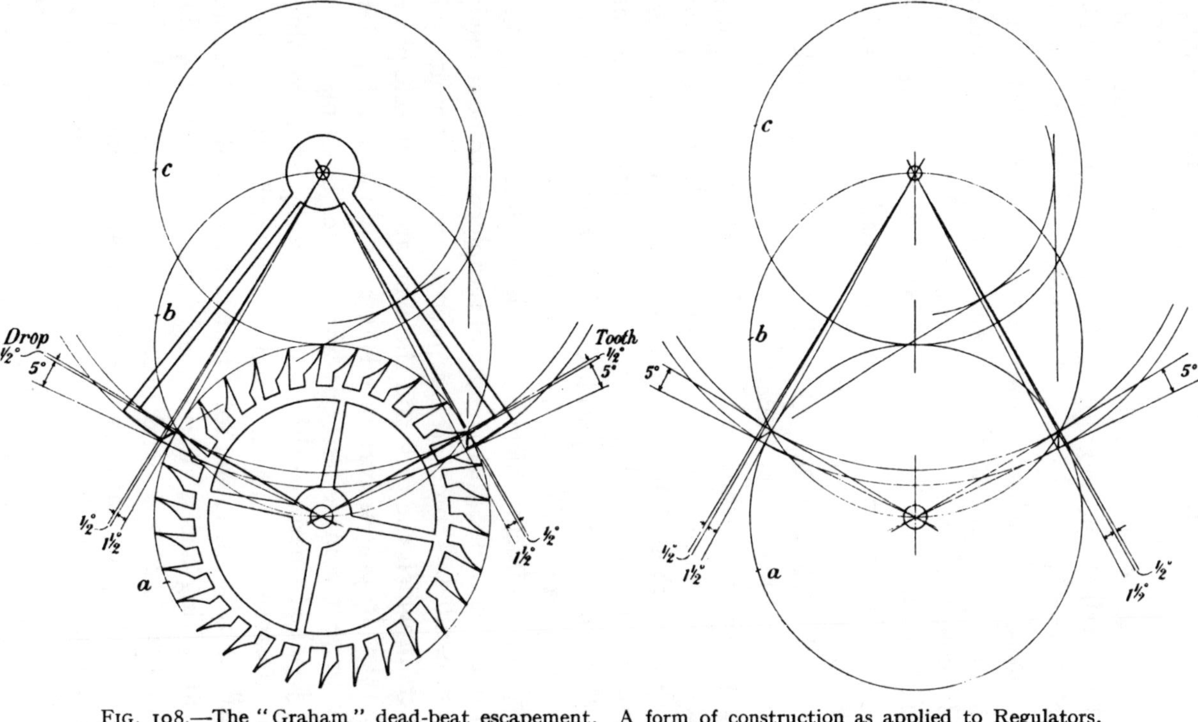

FIG. 108.—The "Graham" dead-beat escapement. A form of construction as applied to Regulators. *Left*: Complete outline. *Right*: Working construction as scribed on templet.

In proceeding with the actual making of this escapement the closest possible accuracy of workmanship is of the utmost importance. The holes of the pallet and escape arbors should all be jewelled and, excepting for the front escape hole which carries the long seconds arbor, they should be on endstones. The wheel requires to be most carefully cut, with fine tips to the teeth. It is usual to scribe an outline of the construction, as shown on the right hand side of the figure, on a flat metal templet with studs at the two centres on which to mount the wheel and pallets temporarily. The action may be checked by this means from time to time in the course of filing the pallets. Makers of regulators in the past filed up the pallets from a steel forging, having first scribed an outline very carefully on the faces. They relied then simply upon eyesight and skill in using the file, leaving them full, and gradually grinding down afterwards with the polisher. In first-class work the acting faces of the pallets are jewelled in sapphire, in which case the steel part is left soft. Nevertheless, so imperative is it to obtain uniform thickness of the pallets and equality of the pallet angles, that the use of templets is to be strongly recommended. Some are prejudiced against "built-up" pallets in which the pads and the arms are not in one piece, yet they do possess certain advantages. The solid pallets may present a more æsthetic appearance, but unless they are jewelled there is the risk of warping in hardening. The greatest accuracy is obtained through grinding and polishing in a lathe, and solid pallets have the disadvantage of not giving access to the inside face of the exit pallet otherwise than by a hand process with a shaped polisher. Detachable pads can, however, be made from a turned ring of steel, and be accurately ground and polished on all sides. Templets to carry and guide the work can be made to fix on the slide rest of the lathe, and then the faces are ground and polished with a stone and lap running perfectly flat in the lathe chuck.

By this means refinement of action, dead drop of the hand on the seconds and accuracy of rate are brought within the range of possibility, always remembering the vital necessity of good workmanship in the train as well as in the escapement, and of ensuring that an extremely firm fixing is provided for the clock in its final position.

Gearing.

(*See pages* 63, 64.)

It should be mentioned that whilst the usual proportion in watch and clock pinions for tooth and space is $\frac{1}{3} : \frac{2}{3}$, this does not apply to pinions of more than 10 leaves, when the proportion becomes $\frac{2}{5} : \frac{3}{5}$. English marine chronometer and regulator pinions may have even thicker leaves than this, in order to minimise shake and contribute to steadier running, less friction and fewer inequalities. In these instances the width of the wheel tooth is left rather less than that of the space, in the proportion of 4 : 5 or 9 : 10, measured on the pitch circle.

In calculations for determining sizes of wheels and pinions, and for the choice of suitable cutters, the basis of circular pitch (page 60) is not altogether convenient in practice, particularly if centres are fixed. Circular pitch is a measurement in inches (or millimetres) of the arc of the circumference represented by the actual pitch. If, however, the ratio is taken of the number of teeth in the wheel and the diameter in inches of the pitch circle, a factor is obtained called "diametral pitch," which is equivalent to the number of teeth per inch of the pitch circle diameter. This bears the same relationship to the circular pitch as the diameter does to the circumference, so that it can be expressed thus :—

$$\text{diametral pitch} = \frac{\text{number of teeth}}{\text{pitch diameter}} \text{ or } \frac{\pi}{\text{circular pitch}}$$

$$\text{and, circular pitch} = \frac{\pi}{\text{diametral pitch}}.$$

Taking as an example a clock wheel of 60 with a pitch diameter of 1·5 inches, then the diametral pitch is

$$\frac{60}{1\cdot5} = 40$$

$$\text{and the circular pitch} = \frac{3\cdot1416}{40} = 0\cdot07854 \text{ inch.}$$

Upon this basis most English and American makers of cutters calculate their tables, but the Continental makers usually adopt another factor—that of " module," which is considered more convenient than diametral pitch for sizing small gearing, as for watch work. The module (M), which has become almost exclusively a metric term, is given by—

$$\frac{\text{pitch diameter in millimetres}}{\text{number of teeth}}.$$

This is also equivalent to $\frac{\text{circular pitch in millimetres}}{\pi}$.

Thus a wheel of 60 with a pitch diameter of 15 mm. would have a module of $\frac{15}{60} = 0\cdot25$. Taking 1 inch as equal to 25·4 mm., the diametral pitch is $\frac{25\cdot4}{\text{module}}$, or to convert diametral pitch into module, $M = \frac{25\cdot4}{\text{d.p.}}$, so that the wheel with a module of 0·25 has a diametral pitch of $\frac{25\cdot4}{0\cdot25} = 101\cdot6$.

The distance of centres of a pair of wheels equals the sum of the respective pitch radii. Since pitch diameter is equal to the number of teeth × module, then the distance of centres

$$= \frac{\text{sum of teeth in the two wheels}}{2} \times \text{module} \quad . \quad . \quad . \quad (1)$$

Or, as the pitch diameter is equal to $\frac{\text{number of teeth}}{\text{diametral pitch}}$, then the distance of centres is also

$$= \frac{\frac{1}{2}\,(\text{sum of the teeth})}{\text{diametral pitch}} \quad . \quad . \quad . \quad . \quad . \quad (2)$$

Simply for the purpose of comparing the two methods, let it be assumed that a wheel and pinion of 60 and 10 have a module of 0·25 or a diametral pitch of 101·6, then the distance of centres can be arrived at thus,

$$(1) \quad \frac{60 + 10}{2} \times 0\cdot25 = 8\cdot75 \text{ mm.}$$

$$(2) \quad \frac{\frac{1}{2}(60 + 10)}{101\cdot6} = 0\cdot3445 \text{ inch } (= 8\cdot75 \text{ mm.}).$$

N.B.—In the formula at the top of page 64 the doubling of the addenda has inadvertently been omitted. The semi-circular roundings have addenda one-sixth of the circular pitch, so that the full diameter is :

$$\text{pitch diameter} + 2\left(\frac{\text{circular pitch}}{6}\right) \text{ or } \left(\text{p.d.} + \frac{\text{c.p.}}{3}\right).$$

Similarly the epicycloidal case may be treated thus : (p.d. + $\frac{2}{3}$ c.p.), but this for practical purposes must be considered only in an approximate sense subject to certain variations.

Involute Gearing.

(*See page* 64.)

At the close of the paragraph on involute gearing mention is made of the uses to which this form of tooth is applicable. It should be added, however, that in recent years the latitude offered in the matter of depth, coupled with the fact that involute teeth can be readily cut by the "generation" process, has led to the extensive use of this gearing in mass-produced clocks, both English and foreign, as well as by some watch manufacturers of cheap grades.

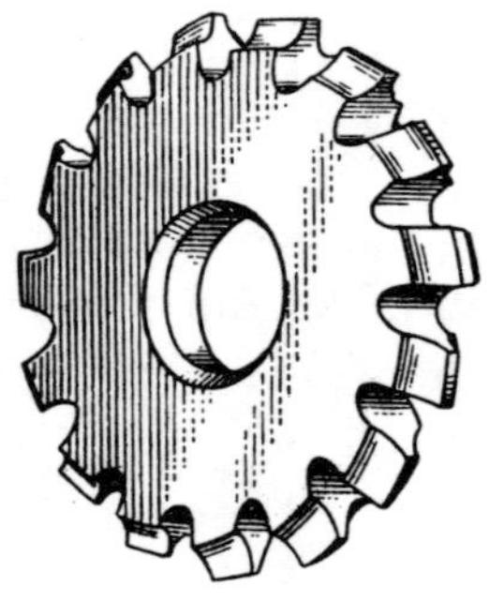

FIG. 109.
Revolving Fly Cutter, for milling involute teeth.

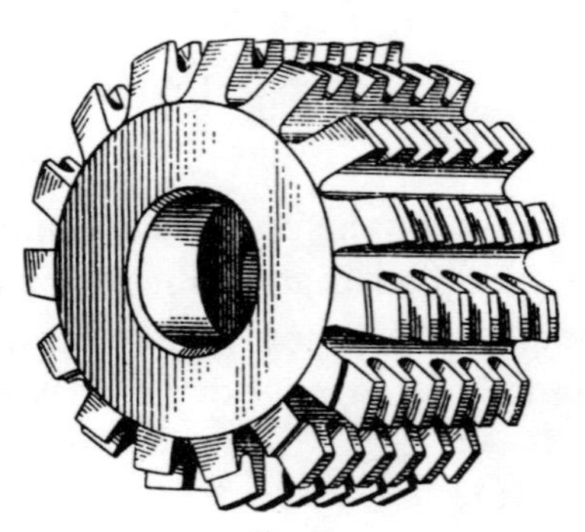

FIG. 110.
Hobbing Cutter for "generating" involute teeth.

The ordinary process of cutting wheel teeth is by milling out a space at a time, either from a single blank or a stack of blanks, indexing the spaces one by one. The revolving cutter (Fig. 109) is moved in a plane at right angles to the blanks. By this means a large range of cutters is needed to suit varying requirements, and although a good cutter can produce very smooth teeth, errors due to warping in hardening and deviations from theoretical accuracy in the actual form, except for the lowest number of each range, make only an approximate degree of precision possible. Again, even with automatic indexing the rate of production is not at all rapid. It is said that the "working time" of a machine is only two-thirds or three-quarters of the "running time."

In the "generation" process, the cutter for involute teeth is a hob or worm with a straight-sided thread, such that the angle of the thread is equal to the pressure angle, and they are ground to shape as accurately as possible after hardening. These hobbing cutters are simply "worms" of six to ten turns (Fig. 110), produced in a screw-cutting lathe and then cross-cut longitudinally with about fifteen indentations, in order to provide a number of cutting faces. In a cutter having 6 threads and 15 cross-cuts, there are thus as many as 90 cutting faces, and, owing to the large number, resharpening is not necessary as frequently as in the case of ordinary milling cutters. Furthermore, the cutter retains its true section, despite grinding until it is worn out. The ordinary backed out cutter changes both size and form with successive sharpenings owing to clearance.

The actual generation arises from the combined effect of the rotation of both cutter and blank, each about its own axis, together with a downward motion of the cutter at right angles to the blank. The blank or stack of blanks has to revolve at the correct speed whilst the hobbing cutter is fed through them, after the manner of an ordinary worm gearing (Fig. 111), the worm engaging with the teeth of the wheel as it cuts them.

The generation principle makes it possible to use one cutter for all numbers of the same pitch, and it is only necessary to set the gears of the machine to the correct ratio—that is, for N teeth in the wheel, N turns of the cutter to 1 of the blank. Taking the example of a 60-tooth wheel, the gear ratio equals 60 turns of the cutter to 1 of the blank. The more teeth in the wheel, the straighter will be the

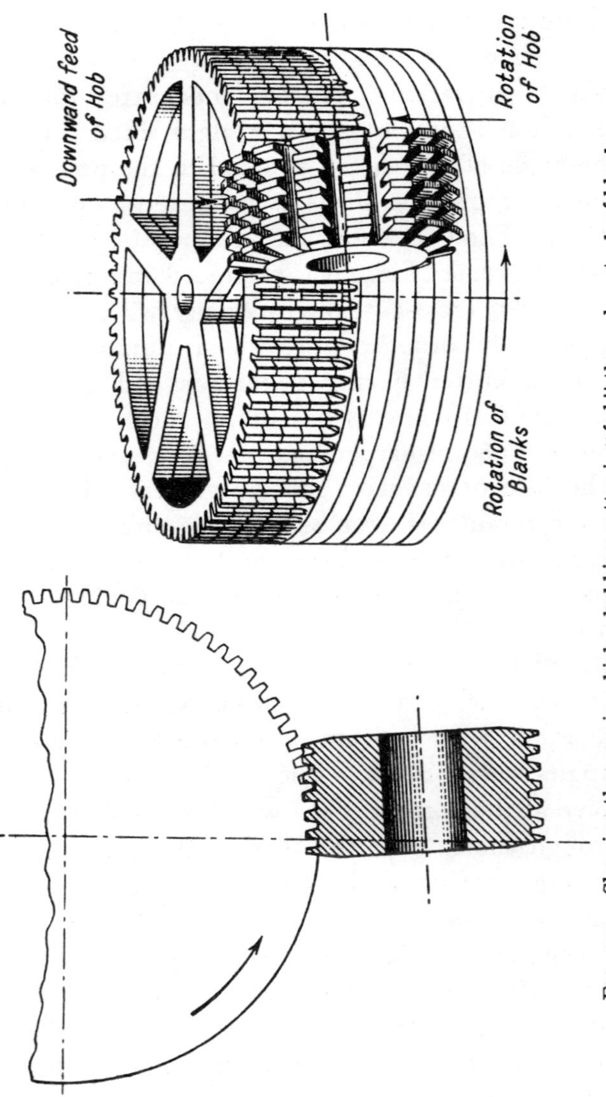

FIG. 111.—Showing the way in which a hobbing cutter is "fed" through a stack of blanks.

profile, whilst in low-numbered wheels or pinions the teeth become bulged and weak at the roots. This can be reduced by making the pressure angle of the teeth larger, to 20 or 25 degrees instead of the more regular 15 or 17·5 degrees. Increasing the pressure angle, however, increases also the tendency of wheels to force one another apart, though up to 17·5 degrees this is not regarded as detrimental, which it is at 20 degrees or more. The pressure angle of any pair of wheels and pinions must be the same—that is to say, if a pinion is made to the pressure angle of 25 degrees, the wheel gearing with it must have teeth of the same pressure angle. It is, however, important to note that as the path of the cutter teeth does not lie in a plane, but in a spiral or screw, there is interference between the cutter and the sides of the teeth being formed. This is more pronounced in thick wheels, for which the method is not suitable. Neither is it suitable for very fine pitches, and there is always roughness of the sides and acting faces of the teeth. Interference is minimised by inclining the axis of the hobbing cutter so that the cutting edges which are spiral in relation to the axis approach the plane of the blanks at right angles, and also by using hobs of large diameter so that the peripheral path of the cutting edges deviates as little as possible from a straight line. In the case of pinions, traces of interference are removed by subsequent polishing.

The small number of cutters required and rapidity of production render the process far less costly than the milling method, whilst the theoretical accuracy of the curves is particularly good. One cutter for each pitch is all that is needed for gears of any number from 12 upwards. In the milling process a range of 8 cutters for involute and 24 for epicycloidal teeth are required for numbers from 12 to 300 of any one pitch, and then only one number of teeth in each range is of correct theoretical formation. For instance, a cutter suitable for 35 to 54 teeth would only give approximate accuracy of form at the lowest number of the range—namely, 35 teeth.

As mentioned formerly, strength of teeth is an advantageous feature of involute gearing and, moreover, the depths do not require to be pitched with the degree of accuracy demanded in the case of epicycloidal teeth. Hence they are far more suitable for manipulation by unskilled labour in the assembling of mass-production clocks and similar mechanisms. They possess the further advantage of running equally well as drivers or followers, which is not general with epicycloidal gearing. Wheels with involute teeth of coarse pitch can also be produced in press tools, and the process is suitable for very cheap clocks, such as alarms. The punches and dies required can be made robust enough and without much difficulty, owing to the sloping flanks of the teeth, which again cannot be done in the case of epicycloidal teeth, with points and radial flanks. This method is sometimes adopted for better work when the pressed teeth are then rectified afterwards with a cutter.

The great importance of accuracy of profile both in tooth and cutter has made the process of verification an almost vital necessity in factories, particularly for the purpose of constructing a correct " master tool " which is used for finishing the curved sides of cutters. This is effected by means of an optical lantern through which an enlarged image of the profile of the teeth or cutter, as the case may be, is projected upon a screen. By varying the distance of the screen, the image is made to register over a scale drawing of the correct geometrical curve. Corrections may then be made to the form of the cutter, making it correspond very closely to the theoretical curvature, which is particularly valuable where fine pitches are concerned. These advantages will be readily appreciated from what has already been said of the importance of ensuring smooth engagement between toothed wheels, not only to minimise friction, but also to maintain steady performance and long endurance.

Origin of the Fusee.

(*See page* 82.)

In dealing with Fusees in the first edition of this book, reference was made to the evidence presented at that time to the possibility of the clock of Philip the Good of Burgundy demonstrating an earlier use of the fusee than the established example by Jakob Zech. As stated, the assertions were based on the excellent description of the clock by the late Dr. Bassermann-Jordan, but there seem, nevertheless, good grounds for questioning the assumption that the fusees and barrels in this clock do really form a part of the original mechanism. This point must, therefore, remain a matter of doubt, and the clock of Jakob Zech (or Jacob, the Czech) which is in the possession of the Society of Antiquaries in London deserves to retain the honour of being the earliest authentic owner of a fusee.

Balance-spring Alloys.

(*See pages* 163, 164.)

The promising characteristics of " Elinvar " as an alloy for balance-springs have been somewhat outshone by the recent appearance of another known as " Nivarox." Composed of iron and nickel with the addition of a proportion of beryllium, this alloy is the outcome of researches made by R. Straumann, technical director of the Fabriques de montres Thommen, S.A., in co-operation with the

Heraeus Vacuumschmelze, G.m.b.H. An account of this alloy was published in the *Journal Suisse d'Horlogerie* (July–August, 1936).

The main feature claimed for this discovery is that not only does nivarox possess the necessary thermo-elastic properties, but also, according to variations of composition, it responds in a compensatory way to magnetic influences, with the further advantage of not causing a falling off in the vibrations of the balance as experienced in the case of elinvar.

The ferro-nickel alloy mentioned on page 164 was the work of Paul Perret, who devoted himself considerably to the possibility of producing a naturally compensated balance-spring. Although this spring when applied to a mono-metallic balance gave a much better rate than a steel spring, compensation was far from perfect. Many made use of it, at the time it was introduced, to advantage in the former English lever watches with plain balances. Dr. Guillaume improved Perret's alloy by the addition of molybdenum, chromium and carbon, whereby a greater limit of elasticity was obtained and a better performance in the middle temperatures. Then he introduced his famous elinvar, which has proved most satisfactory from the standpoint of compensation, but compared with tempered steel it has not the same vitality for maintaining a good action in the vibrations of the balance.

In order to provide elinvar with the greatest degree of elasticity, titanium has been added to the alloy, but the result has not corrected the weakness of the vibrations, nor altered the tendency to magnetism. Straumann has sought to overcome these deficiences by using beryllium in place of carbon in the alloy of iron and steel, and on this basis the nivarox alloys have been evolved. The addition of beryllium in conjunction with molybdenum, tungsten and chromium makes it possible to reduce the internal molecular friction of nivarox springs in such a way that the vibrations are said to show even greater activity than with a steel spring, whilst the susceptibility to magnetic influence can be practically abolished, corresponding to the conditions existing in non-ferrous springs. Their behaviour in fluctuating temperatures is also said to have reached a degree of perfection which is capable of meeting every contingency.

The temperature-compensating action of nivarox depends upon the combined effects of certain internal stresses or forces, which are partly mechanical and partly of a magnetic nature. It is supposed that the internal magnetic forces of steel, nickel and their alloys, which are parallel in small particles of the order of 0·01 mm. in size, vary so much in direction from one particle to another that the minute magnetic fields neutralise each other, and so no evidence of magnetism is apparent on the exterior of the body. These internal forces, however, influence the modulus of elasticity of the substance. In order that the modulus of elasticity should remain constant with varying temperatures, it is necessary that the mechanical and magnetic forces should be affected to an equal degree, but in opposite directions, so that the balance between them is undisturbed.

By a process of thermal treatment it is possible to control the value of the mechanical stresses and to determine the thermo-elastic coefficient in the making of nivarox springs, which can be graded according to requirements. Under different categories there are springs almost non-magnetic, suitable for cheap mass-production watches; others quite non-magnetic and corrosion-resisting, for use with the bi-metallic cut balances of pocket and marine chronometers, whilst, to meet the needs of the very best watches and chronometers, there are some entirely noncorrosive and capable of reducing the middle-temperature to zero or a negative value.

Marine Chronometers.

(*See page* 243.)

It should be remarked that the wording of the Government's offer, which was successfully challenged by John Harrison, actually called for " a method " by means of which longitude could be determined at sea, not as stated, " an instrument."

Kullberg's Auxiliary Balance.

(*See page* 252.)

The description at the foot of the page of Fig. 105 is inclined to mislead. It should read thus : This auxiliary rim, A, A_1, is cut through at C, C_1, and being thicker, offers resistance to the outward movement of the main rim under the influence of decreasing (colder) temperatures, by means of the blocks or kneed pieces, B, B_1, which impinge against the banking screws, D, D_1.

REPRINTED BY LITHOGRAPHY IN GREAT BRITAIN
BY JARROLD AND SONS LIMITED, NORWICH